T0032929

Praise for

Bootstrapped

Kirkus Reviews Best of 2023: Nonfiction

"Alissa Quart is an essential chronicler of American inequality. . . . We need her work more than ever."
—*Literary Hub*

"A provocative, important repudiation of gig economy capitalism that proposes utopian rather than dystopian solutions."
—*Kirkus Reviews*
(**starred review**)

"In *Bootstrapped*, the author (*Squeezed*) and executive director of the Economic Hardship Reporting Project brilliantly debunks the American fantasy of self-reliance. . . . Her book is a thoughtful, nuanced examination of our 'self-punishing individualism.'"
—*New York Journal of Books*

"As the heir apparent to the late, great Barbara Ehrenreich, Alissa Quart takes on the American myth of the self-made man. . . . A must-read for understanding our moment and a fierce wake-up call."
—**Beth Macy,**
author of *Dopesick*

"The reporting and storytelling here are incredible; my jaw literally dropped several times. . . . Ultimately, this book left me with much-needed hope, resolve, and curiosity about all the things we make possible together."
—**Jenny Odell, author of *How to Do Nothing***

"In this elegant and incisive book, Quart challenges us to see a through line of self-creation from Thoreau to Reagan, weaving together stylish analysis and evocative reporting. Quart is a fantastically entertaining literary class warrior." —Zephyr Teachout, author of *Corruption in America*

"At a time when many are struggling, Quart digs into the idea of the American Dream and asks how we can truly help one another thrive." —*People*

"Quart's case studies of organizers and rank and file in different movements are prescriptions for systemic changes, while her urgent case for reframing how we think of ourselves and our society presents an individual-scale project for all of us." —Cory Doctorow, *Pluralistic*

"Reading *Bootstrapped* is a good reminder of how much we depend on one another. Applying the ideas of this book to an academic context might help us understand the structures that stand in the way of success for everyone within our university communities." —Inside Higher Ed

"If all *Bootstrapped* aimed to do was expose the hypocrisy of those who promote the myth of total self-reliance, it would still be a well-written and valuable contribution. But Quart's book has a larger point to make: there is simply no such thing as true independence within the human condition."
—*Jacobin*

Bootstrapped

LIBERATING OURSELVES
from the AMERICAN DREAM

ALISSA QUART

An Imprint of HarperCollins*Publishers*

For Barbara Ehrenreich and my family

BOOTSTRAPPED. Copyright © 2023 by Alissa Quart. All rights reserved. Printed in the United States of America. No part of this book may be used or reproduced in any manner whatsoever without written permission except in the case of brief quotations embodied in critical articles and reviews. For information, address HarperCollins Publishers, 195 Broadway, New York, NY 10007.

HarperCollins books may be purchased for educational, business, or sales promotional use. For information, please email the Special Markets Department at SPsales@harpercollins.com.

Ecco® and HarperCollins® are trademarks of HarperCollins Publishers.

A hardcover edition of this book was published in 2023 by Ecco, an imprint of HarperCollins Publishers.

FIRST ECCO PAPERBACK EDITION PUBLISHED 2024

Designed by Angela Boutin

Library of Congress Cataloging-in-Publication Data has been applied for.

ISBN 978-0-06-302801-2 (pbk.)

24 25 26 27 28 LBC 5 4 3 2 1

Contents

Preface: Forget Self-Reliance

Obsessed, bewildered/By the shipwreck/Of the singular/
We have chosen the meaning/Of being numerous.

—GEORGE OPPEN

I RECEIVE MESSAGES on a routine basis from strangers about how the poor are responsible for their own poverty. Those who are economically on the edge, they write, just need "to pull themselves up by their bootstraps."

The complaints come in the form of emails and comments regarding the supposed bad choices of the financially unstable. These emailers find fault with the underresourced for ostensibly choosing to be single mothers or not saving themselves for marriages to good providers or for being evicted for being "illegal immigrants" or even for trying to continue to work as journalists. They wag their fingers at the indigent for having college or graduate school debt and for not getting adequate job retraining, not seeing this as a paradox. Others lecture the economically unstable for putatively wallowing in their condition.

It was not happenstance that I was the recipient: I was getting this stream of invective from news consumers because I run a journalism nonprofit called the Economic Hardship Reporting Project, which is devoted to covering income inequality and poverty, and I have also spent much of the last eight years reporting on these matters. Some of my organization's writers have experienced homelessness or eviction or they have had to watch family pets die because they couldn't afford vet bills; some of my sources have experienced similar setbacks.

And as a result of editing, writing, and publishing these stories, I have learned more about this toxic ideology. It's classist, sure, but it's also a prime example of how our country's most unprotected, its poorest citizens, are routinely and publicly shamed, a kind of nation-wide bullying.

The letter writers and commenters were not just apt to critique others for a presumed shortage of self-discipline. These armchair critics also did the inverse, glorifying their own supposedly independent lives in notes and posts and on call-in radio shows. One informed me that he had thrived even though his own parents had only GEDs. Another's mom was only a teenager when they were born, but they now earned six figures. The letters and comments often concluded with a celebration of how my correspondent had lived without government or other assistance or supported a child or children through their household earnings alone, including the incomes of their (very heterosexual) husbands, if they were women. They also loved to note that they themselves had "managed to save some money, had a car, a TV, and food." As one writes to me, "I wish younger people would stop trying to blame others for their problems and look within. We are all products of our choices, and unfortunately, sometimes we just have to live with the consequences!" Another remarks, "You are responsible for everything you do in life . . . you are talking about giving people an income, talking about free health care. Your health is your responsibility . . . I drive a 2011 pickup truck because I do not need to buy a new one." Or: "In your view, it seems to be OK to live beyond one's means without regard to the financial consequences, because the government (or someone) should pick up the slack."

In my childhood, we called these sorts of sneering voices the "peanut gallery"—back then, I thought that meant you literally threw empty shells at people onstage, from a distance. And it still seems appropriate, because they are the crowds who refuse to take others' expression of their economic straits seriously and throw detritus at anyone who dares to be open about their financial realities. The gallery likes to minimize the level of the other person's suffering or the effort they put in. And they enjoy the supposed moral fragilities of those who are economically vulnerable. They sneer: Didn't they *choose* to send their children to college, perhaps taking on educational debt? And why had they "chosen" to have children in the first place? And how dare they own a modular couch or a flat-screen television? The peanut gallery might ask these questions of others, as if by ostensibly having made "better" choices for themselves they would feel protected from future fragility or illness or other human losses. I have come to realize that these sentiments don't run contrary to the American Dream. Instead, they express the dream at its worst.

I call the way these folks chase this stereotyped version of success "bootstrapping." It's a shorthand term I am using to describe this every-man-for-themselves individualism. It also defines how this mentality affects those who can't make it, in the conventional sense, and have internalized the bias against them.

Bootstrapping, that is "pulling ourselves up by our bootstraps," also contains many of the contemporary tropes and archetypes, like "girlbosses," the updated millennial version of the self-made corporate woman, and "side hustling." And as you will see in this book, it's no accident that many of the trendy idioms describing our economic state may sound appealing and cool, but they most often simply describe the anxious experience of being financially on one's own. These phrases are part of what it means to dwell in a bootstraps society. In addition, there are phenomena that are unidentified yet ubiquitous. I have named as many as I could in this book, among them the "dystopian social safety net," a term to describe social programs or crowdfunding arrangements that should not *have* to exist yet we rely on heavily. That the dystopian social safety net is there at all is due to the fact that we live in a nation that demands we be impossibly self-reliant. Much of this book is devoted to better understanding these social constructs, conceptions that serve to cover up how little societal care we receive.

Those who subscribe to the fantasy of self-sufficiency believe that people can and must make it on their own—and studies like a recent one by the nonprofit organization the Moving Up Media Lab have found that indeed most Americans think success is something one achieves alone. And people who believe this are also more likely to judge others who are in precarious financial situations and are more inclined to deny the role of inborn advantages or any outside help they received. Some examples of this blame-iness include a 2020 Pew study that revealed that 42 percent of Republicans say those who are poor are indigent because they have not worked as hard as most others. In addition, 60 percent of Republicans agreed with the statement "People get stuck in poverty primarily because they make bad decisions or lack the ambition to do better in life," according to a 2019 Center for American Progress survey. (Others supported statements like "If everyone tries hard, everyone can get rich.")

This mindset is so pervasive that it leads to self-blame, with people wondering what they may have "done wrong" to wind up in their circumstances; often the reason was being born to a family without adequate savings or having accrued educational debt due to the higher learning they were told by their teachers and political leaders to acquire. Those left out of the feeding frenzy of self-betterment fault themselves. According to a 2015 Oxford University and Joseph Rowntree Foundation study, poorer people tend to experience a "negative self-stereotyping" effect, absorbing the media clichés and considering themselves "low in competence," and even flawed at the root.

Over the last eight years I have seen how the stark economic gradient and the faith in a certain version of the American Dream fed on each other in strange and dark ways. It can be a form of what Lauren Berlant, a University of Chicago cultural theorist, calls "cruel optimism," where your own desire prevents you from thriving. In other words, our country's ideal of happiness is a fantastic and impossible pursuit, and one full of stress. It is bad for us, a veritable deadly nightshade.

In my reporting on those chasing this dream of self-sufficiency, I have found some have become rightfully cynical about the possibility of individual triumph. Given our unstable jobs, unpayable college and medical debt loads, and the fact that we dwell within a byzantine financial and tax system that benefits the wealthiest, skepticism as an overall stance makes perfect sense. Others remain uncritically entranced by the anthem of solo achievement.

In addition to this contaminated binary, though, during the pandemic I saw a third impulse afoot. There was a new level of economic empathy and community-mindedness embodied by small-scale democratic workplaces and novel citizen altruism and activism. When COVID raised its ugly crown, a more lambent reality than the one represented by these commenters was now on display daily. We were suddenly being made aware of how, and on whom, we were dependent: the grocery store clerks who were ringing up wipes and twelve-packs of paper towels; the workers delivering medications to the elderly, or at least those who could afford such a service; the postal workers collecting our ballots and bringing to the unemployed the last of their paychecks; the cleaners who made subways smell like antiseptic and lemons; the strawberry pickers of the Central Valley of California who ensured that fruit would arrive in supermarkets; the seamstresses who once stitched designer dresses now sewing face masks; the spirits' manufacturers who reinvented themselves as hand sanitizer mixologists; the theater arts directors who built pop-up hospitals; those laboring at pharmaceutical plants that keep us on cholesterol medication and Ambien; the health workers heading home in their jewel-toned scrubs, as eye-catching and meaningful as navy whites or priests' frocks once were.

The pandemic has, whether we realize it or not, drenched us in political opportunity and revelation, both showing us once again that we exist within the antique sociological construct of "organic solidarity." This kind of cohesion is not societal togetherness forged of common beliefs and values—in a place as various as the United States, that would be hard to come by. Rather, organic solidarity means depending on one another for practical reasons, because our work is specialized and complementary. The farmer needs the truck driver who needs the schoolteacher who needs the students. Parents need the caregivers. We need reporters to show us the truth, whether we still subscribe to newspapers they work for or not. (Anyone who thinks they are truly self-made should call their mother.)

We might achieve this welcome state of personal and societal interdependence if we rejected the individualism con. I won't say that I am always the best at achieving this myself. I was raised in this country. The bootstrapping story makes me overwork in the name of achievement, even as, paradoxically, all this isolated work is geared toward greater solidarity.

I believe, however, that there is another story line out there, a better tale, one that isn't simply a lonely, faltering, and often doomed trudge toward personal financial victory. If this is the American Dream, we must collectively wake up.

I

CREATORS OF THE AMERICAN DREAM

1

The Backstory

There are in the world no such men as self-made men.
The term implies an individual independence of the
past and present which can never exist.

—FREDERICK DOUGLASS, "SELF-MADE MEN"

THE CLASSIC LEATHER boot has had many names over the years—lace-up, cowboy, congress, pale rider. To get your work boots on your feet, you'd stand up and grab two small leather flaps on the sides, known as bootstraps, and pull the boot up.

From this everyday activity, the nineteenth-century idiom "to pull yourself up by your bootstraps" was born—and with it, a torturous myth that true success means getting ahead only on your own energy and steam, without help from your family, government, or community. In short, this idiom represents how we've been told we need to make it as Americans.

The bootstrapping story is the tale of our individualism, shading into a brittle self-sufficiency. It is the conceit that we must take care of ourselves wholly, that we must make our own luck. To claim that we have acted with utter independence usually means we are denying something. In truth, to bootstrap is to disregard or erase the roles of our parents, teachers, or caretakers as well as the roles of wealth, gender, race, inherited property, and a whole cache of related opportunities.

The myth of the self-made man (and far more occasionally the self-made woman) also enforces the pernicious parable of the deserving rich. You can hear the reverb of this societal delusion when people say of their financial success, "I did it alone," denying the web of relatedness all of us dwell in. The bootstrapper yelps, instead, "I wrote my own book" or "I started that cargo company"—when in fact she hired a ghostwriter or he borrowed a hundred grand from a school friend to launch his first ship. The individualist fantasy also shapes some voters' choice in politicians, as they are drawn most to those who insist that they have succeeded entirely by themselves.

Bootstrapping can induce self-blame as well, as I have mentioned. The political pollster and thinker Anat Shenker-Osorio observed to me that an inordinate number of the middle-class Americans in her focus groups believed that "if they didn't make it, it's because of themselves" and expressed self-loathing around any perceived failures or setbacks they had experienced.

The cult of individualism has led generations to lacerate themselves.

But while the game may be fixed by others, we still often have a nagging sense that our failure is ours alone. It's like being a gaslit partner in a long, abusive marriage, told one thing again and again until we believe it. That one thing is that we are the sole creators of our own destinies, and if we haven't quite managed to achieve, what value do we have? If self-reliance is the ultimate virtue, those who can't or won't manage this circumstance are encouraged to find themselves wanting. "Bootstrapping" rests on a notion of personal responsibility, as well as one where the wealthy are deserving of their riches and the poor and the strapped middle class deserve to live on the edge, where they can compare bad credit scores and battle to get their children into college. The bootstrapping myth drops the blame for inequality in our laps, while our flawed systems get off scot-free.

When we embrace this self-obsessed stance, though, we are ignoring our own biology; we are a social species wired against isolation. While triumphant aloneness has been enshrined as a value in many an American biography—and every rock-climbing documentary—in the real world, such self-sufficiency will set off stress alarms in our bodies, because we are animals programmed to seek connection with and empathy for others. In fact, these stress alarms can be turned off and calmed only by connection and belonging; they are also the key to our ultimate happiness and deep-seated success. Many studies have shown that friendships and our sense of community are what really help us relieve anxiety and literally make our blood pressure decrease. The narrative of walled-off striving goes against our biological need for the collective, one that is built into our brains. And as the economist Joseph Stiglitz reminded me in a conversation we had, being self-made was in itself a clear biological as well as social lie: "Biologically you can't be self-made, as no one is: you always begin life with gifts from your parents . . . or liabilities."

═══

When the concept of pulling yourself up by your bootstraps was first advanced in 1834, it was understood as surreal, intended to be seen as an outlandish act—how could anyone pull up their boots to lift their own bodies? As the broadsheet *Working Man's Advocate* puts it, tartly, in an article about a local inventor named Nimrod Murphree who won patents on his agricultural devices: "It is conjectured that Mr. Murphree will now be enabled to hand himself over the Cumberland river or a barn yard fence by the straps of his boots." (The man was being mocked.) A little later, the February 14, 1843, edition of the *Southport Telegraph* out of Southport, Wisconsin, had an item in which it quotes the *Racine Advocate*: "the Governor must be trying to pull himself up [by] the bootstraps."

Part of the interest in bootstraps to begin with was how hard boots were to take on and off in the nineteenth century, far from the casual slip-on clogs of today. Boots were so difficult to get on that they required mechanical devices, which wealthier people used to get their footwear on and off. The even richer needed servants to help them slide their boots onto their feet.

The absurdist use of the idiom continued long after Murphree's name appeared in print. For instance, a critic for *The Dial* magazine in 1860 deploys the phrase to indicate a mental paradox. "The attempt of the mind to analyze itself [is] an effort analogous to one who would lift himself by his own bootstraps," they write. "Metaphysics is holding your mind in your mental teeth," they continue. (When I read this, I try to hold my mind in my "mental teeth.")

Throughout the nineteenth century, the concept of bootstrapping still retained some of its ludicrousness in the public imagination. While it was understood as a cartoon, it was also absorbed as a real thing and an aspiration. It matched another phrase that came into fashion in that period, "self-made man," as popularized by the Kentucky politician Henry Clay in 1832. It was embraced in a zeitgeist of grandiose impresarios looking to make a buck.

Both the self-made man and bootstrapping implied the individual's capacity for scrappiness but also domination. These words wormed their way into millions of people's minds. Eventually, somehow, "pulling yourself up by your bootstraps" became laudatory. The concept of the lone individual achieving, unaided by others, was enshrined. The leitmotif of Man Alone was celebrated in the work of nineteenth- and twentieth-century authors like Horatio Alger and many others. This kind of—let's face it, male—individualism was hardly as virtuous as it was cracked up to be.

It has fed into the extreme rhetoric and actions of everyone from robber barons of yore to Reagan Republicans.

"The size of the federal budget is not an appropriate barometer of social conscience," Ronald Reagan said, as he used his metaphorical buzz saw to take apart welfare, coming up with a whole language to demean those who were dependent on state monies, including "welfare queens." In 1971, Reagan, then the governor of California, called such social aid "a cancer eating at our vitals." The real-life welfare queen was "a woman in Chicago," as Reagan put it, who had actually committed public-assistance crimes. She was only one criminal, yet she became the mythic scapegoat in antipoor rhetoric. In fact, Reagan's mantra about one midwestern woman became the foundation of his stump speeches in his presidential campaign, starting in 1976.

In the years leading up to Reagan's presidency and during those of his tenure, as the economic historian Pamela Walker Laird and others understand it, to be a self-made success meant you were morally good, and if you had failed to succeed, you were morally corrupt. This ideology of bootstrapping, filtering through American politicians ranging from former president Donald Trump, with his faux self-made mindset, to ultraright Senator Josh Hawley and Tim Boyd, the now-former mayor of Colorado City, Texas. (During a period in 2021 when unusually frigid temperatures ravaged his state, Boyd boomed his bootstrapping views in response to a population that was in some cases freezing to death: "Sink or swim, it's your choice!")

The pulling-yourself-up-by-your-bootstraps ideology also suffuses Silicon Valley companies and punitive local school boards. It is present when an unholy number of people buy into accusatory narratives around other people's instability and form a nasty conclusion: why should taxpayers support the poor? In this equation, those in economic need are somehow always already ethically benighted. It's also part of how many of us think about economic uncertainty. That's why, for me, some of the usefulness of the word in past tense, *bootstrapped*, is that it contains the word *strapped*. After all, being strapped for cash and time, due to income inequality, contingent work, and lack of a strong social net, leads people to strain even harder toward success, often to no avail. *Bootstrapped*'s other meaning—when an entrepreneur starts a company with meager capital—is also part of the appellation's value.

Yet despite the disparities that are built-in in our country, from a young age Americans are taught to blur out the disadvantages that our social differences have created and attribute success to the inherent better character of the winners and then call the whole thing a meritocracy. That's even though our country has shifted the burden of our survival almost entirely onto their backs, preferring the half solutions of big philanthropies or charity such as crowdsourcing efforts like GoFundMe—not a social service provider!—to offering permanent public childcare or a supported health-care system.

The happy news, though, is that our country is large enough to contain rival and antithetical tendencies. As much as individualism dominates, millions in this country have also pushed against the singular and toward its opposite coming together in cooperatives, collectives, and mutual aid societies. Historically, this strain also could be seen in group efforts like barn raisings, where farmers and their friends would build one member of the community's barn for free. Or we might consider the antirenters, who together in the 1840s went on rent strike against their landlords and publicly protested in costumes like something from Bread and Puppet Theater in an effort to obtain the equivalent of DIY affordable housing. (Some scholars have argued that their actions even led to the Homestead Act later in that century.) And during the Great Depression, there were similar collective efforts started by individuals and stemming from the federal government.

While the fairy tale of solo success fails many Americans, there are alternative models that can take that fiction's place, ones rooted in the tenet of interdependence and working together to lift one another up. We might also accept our dependence, permitting and acknowledging societal aid and help from other structures of support.

Call it antibootstrapping. It's an alternative framework to the individualistic one that has separated us and shamed us for nearly two centuries. In this book we'll explore what can be learned from these oppositional efforts. National prosperity requires "community support as well as individual effort," as Laird reminds us in her book *Pull*. These attempts didn't arise only from the radical margins. After all, it was the Reverend Dr. Martin Luther King Jr. who said, "It's a cruel jest to say to a bootless man that he ought to lift himself by his bootstraps. And many Negroes, by the thousands and millions, have been left bootless . . . as the result of a society that deliberately made his color a stigma." When President Barack Obama occupied the White House, he also invoked faith in the collective, underlining that previous generations and the infrastructure they made "helped to create this unbelievable American system that we have."

Similarly, in his 2021 inauguration address, President Joseph R. Biden Jr. gestured at the principles of community-oriented public-spiritedness: "There are some days when we need a hand. There are other days when we're called on to lend one. That is how we must be with one another. And, if we are this way, our country will be stronger, more prosperous, more ready for the future."

Progressive public figures have also shown distaste for the self-made myth, as when they have spoken humbly about their own accomplishments, vanquishing the idea that individual success can be accomplished in a vacuum. "I worked my butt off to get elected against all odds, without any special connections or money," tweeted US Representative Alexandria Ocasio-Cortez, the Queens and Bronx, New York, Democrat. "I worked double shifts and wore through my shoes, outspent 10:1 to get elected." But even with all that hard work, she continued, "it would be narcissistic to pretend I 'bootstrapped' it alone."

In addition, activist and labor groups reflecting this community-first philosophy have grown in power in the midst of the pandemic, from gig worker collectives to volunteer citizens' groups. These citizens' organizations embody the richness that comes from knowing how to rely on one another rather than going it solo.

What is often forgotten in this dialectic of doing it alone versus grassroots solidarity is that there is still yet another way: governmental solutions that far exceed what we can accomplish on our own in self-organized groups. These solutions rest, however, on higher taxes. And that is not an impossible aim either: during one of our nation's greatest eras of growth, the 1950s and '60s, it was not bootstrapping but high taxes that made our country great. In the Eisenhower era, Americans paid a top tax rate of 91 percent. (In 1943, the top tax rate was actually 92 percent.)

———

It was meant to be a boom time when I began this book, a period of low unemployment, of high consumer confidence. Then the pandemic descended on us and accelerated and exposed the faults that were already present in our institutions and systems. For one thing, it revealed our vulnerable labor market—and how thin the veneer of prosperity was for most of us.

Back then, it felt as though our society was cleaved into that old binary that I think of as bootstraps vs. community. For obvious reasons, the least likely people in 2020 came alive to the idea—though perhaps only the notion—of what is sometimes called interdependence. The words *dependence* and *interdependence* were uttered not only by locals volunteering to shop for their elderly neighbors but also by politicians.

One of the most beautiful public expressions of interdependence came from the Reverend Raphael Warnock when he was elected to the US Senate in 2021 after the Georgia runoff. It took the form of a quote from Martin Luther King Jr.: "He said that we are tied in a single garment of destiny," Warnock recalled, "caught up in an inescapable network of mutuality." This reciprocally minded sentiment acknowledges that *self-made* is an oxymoron.

But leaders who explain that those who succeed tend to be born on third base, as the cliché has it, are few and far between. Instead, there is both a fantasy on offer of quick trips to Easy Street on one hand and decades of dour technocratic pronouncements of glacial progress on the other.

In truth, transformation can, on occasion, be rapid. In the last decade, we have sped along policy ideas previously considered radical or obscure. Americans had come together to create cultural shifts like the legalization of gay marriage, gender equity in the workplace, minimum wage hikes, and, yes, even Trump's election.

More recently, Biden fought for a bill that centered on "human infrastructure," a term popularized by Vermont senator Bernie Sanders to describe benefits that help ordinary folks, like welfare programs and other kinds of people-centered offerings. Even though much of it has been opposed at every turn in the Senate, his has been one of the most progressive presidential agendas in recent history, from attempting to address climate change to proposing childcare tax credit and free community college. In the honeymoon period of 2021, at least, perceived pragmatic limits weren't boundaries after all, and perhaps pragmatism itself was recognized as the problem in the first place. Despite reliable pushbacks and setbacks, the start of the Biden administration revealed a new level of public support for items like paid family leave (although the four weeks that were on offer in October 2021 still put the United States nearer to the bottom of the 174 countries that offer paid leave for personal health problems). The pandemic was like an Overton window for a different way of thinking about societal care and interconnectedness, from paid leave to childcare assistance.

Soon enough, the torments of a riven Democratic Party and extrapartisan Republicans returned to haunt our country, with reactionary forces working, with great war chests, to shut these ideas and movements down, and so far, relatively successfully. Nevertheless, that these societal policies and notions have now been brought to the fore and popularized is itself a signal. It potentially gives us renewed strength against the usual hectoring rhetoric that anything universal is "not how the world really works." It can better help us ward off such negations delivered by legislators, one percenters, and Aspen Ideas Festival denizens.

The truth is if we stop being loyal to the status quo, we can accomplish swifter change than we might imagine. Who would have thought the House would pass a spending bill that provides for free community college? Or that we might experience eviction moratoriums beginning in March 2020 and see drug decriminalization bills beginning in November 2020? These sorts of transformations— through measures, initiatives, and spending—would have seemed like something from progressive fan fiction.

When we embrace a value system outside that of "pulling ourselves up by our bootstraps," one that rests on communal strength, these changes are at least possible. With such a realignment of our faith and imagination, majoritarian and permanent programs become not only possible but inevitable. In this singular time, we have a unique opportunity to bring more people to a collectivist mindset and away from the vainglory of billionaires in space. And in the aftermath of the pandemic, we have the unique possibility to rewrite our tired cultural narrative that exalts lonesome achievement and spurious self-sufficiency above all.

With each person who sees our country through that collective lens and puts these values into practice, we move further from the dusty, damaged first person and into the first person plural. That's why, among other things, I consider this book to be radical self-help. Lifting up one another should never be considered weakness. Leaning on one another almost always makes us stronger.

2

Where's Walden?

It is the age of the first person singular.
—RALPH WALDO EMERSON

THE PEBBLY SHORE thrummed with people: a medley of humanity. It was January 2020, before the pandemic made such a thing lethal, on Walden Pond. I was close to the homes of two patron saints of self-reliance, Ralph Waldo Emerson and his best friend, Henry David Thoreau. The two were famous Transcendentalists, that literary and philosophical school of thought that helped define the middle of the nineteenth century. That movement centered around nature and was influenced by Romanticism, with some more radical ideas swirled in. And the two men's preoccupation with the meaning of individualism helped to some extent to define the American way.

But did the high-literary figures of 150 years ago really help define the sentiments of, say, the Silicon Valley rich who live in Walden Monterey, a luxury Bay Area development named after Thoreau's book? Could these thinkers have influenced the bootstrapping Trump voters or even the titans of industry, including the man who invented the computer I am typing this on? Yes, and yes, as did a number of other authors who exulted in self-making, like Horatio Alger.

We are still living within Emerson and Thoreau's notions of self-sufficiency, to some extent.

In order to understand this all better, I visited William Huntting Howell, a specialist in Emerson's work and the thinkers of his era. Huntting Howell argues in a book that early America is not entirely a story of stoic independence. After I met him at the site of the two men's homes, in front of the large white clapboard house where Emerson dwelled, Howell explained that while the man was a Romantic and antislavery, someone who read philosophy and supported elements of women's rights, his writing frequently reflected a stance that in the future could be called "corporate individualism." As Emerson wrote in his famous essay "Self-Reliance": "Do not tell me, as a good man did today, of my obligation to put all poor men in good situations. Are they my poor? I tell thee, thou foolish philanthropist, that I grudge the dollar, the dime, the cent I give to such men . . . I sometimes succumb and give the dollar, it is a wicked dollar, which by and by I shall have the manhood to withhold."

Within Emerson's parsimonious words was the spirit of his time and to a lesser extent ours as well.

It was in the air around Emerson when he wrote, though. In 1832, Kentucky congressman Henry Clay originated the term "self-made men." Clay gave a speech to the US Senate describing the "autonomy of our manufacturers." He was a proponent of what today would be called entrepreneurs but at that time were known as "industrialists." He was known as "Great Harry of the West" in keeping with what he called the "rugged pioneer spirit" that accrued to the self-made man myth he had helped create.

This fervor for self-made men grew in the 1830s and '40s, along with frequent public praise for new enterprises like shipping companies, the start-ups of their time. Self-made men seemed to be the ones behind the new and more efficient steamboats that were all the rage then, on which cargo could be shipped more quickly, cheaply, and widely. The term was also used to describe the new breed of Western entrepreneurs. To further this development, Clay and other politicians endorsed domestically favorable tariffs that would heighten support for American business.

That era marked a break with the recent past. As political philosopher Nancy Fraser and her collaborator Linda Gordon observe, being in a state of "dependency" in the preindustrial period was not the onerous thing it became. It was rather framed as the natural state for so-called societally subordinate people—women, children, or servants. "Depending on" or "hanging from another," they write, was not necessarily a source for humiliation: dependence-shaming started, instead, when being needy was subjected to moral scrutiny. In the nineteenth century, with the rise of factories and steam trains and their promise of wild individual wealth and more staccato societal mobility, there was suddenly, they argue, "the shadow of a powerful anxiety about dependency." What was once considered a normal reliance on others had become "deviant and stigmatized." Now, if one didn't subsist solely on one's own steam (or steamboat!), one deserved contempt. Back then, as now, bosses relied on underpaid workers and husbands relied on wives: invisible labor has a long and tawdry history.

=====

At forty-six, Howell himself is a pure product of New England, so classically northeastern that he is the third William Huntting Howell in his family. As such, he gravitated to Emerson and Thoreau while his peers studied more au courant writers. In fact, Howell first got hooked on the Romantic Transcendentalists before he was even a scholar, when he was just coming of age in the 1990s. He became more fascinated by them in graduate school, in the age of President George W. Bush and the Iraq War, when he suspected them of secretly influencing the whole culture of militarized individualism—the gun-happy sniper of the pro-military film *American Sniper* is all too self-reliant, after all—that defined the early aughts.

Talking to him, I realized how I and so many others had been trained to think of Emerson and Thoreau as high-minded Romantics, as nature-loving idealists and Great Men™ who doled out an ideal of lyrical intellectual self-sufficiency. It was this version of the absolute autonomy that I found most hypnotic, because they presented their stance as independence of *the mind* rather than the bank account.

In truth, Thoreau's life did not match his rhetoric of solitude and singular mastery. Thoreau may be famous for his isolation, living and turning his face "more exclusively than ever to the woods," in order to "live deliberately, to front only the essential facts of life, and see if I could not learn what it had to teach." For the two years he dwelled on Walden Pond and composed his book, he sold himself to the public as utterly self-reliant.

While Thoreau played at being the ultimate Man Alone, in fact he leaned quite heavily on others. (Spoiler alert: I am going to commit a cardinal sin of twentieth-century literary theory by gazing at these writers through biographical lenses. So be it.) Thoreau used Emerson's library, and as once a collegiate fanboy of Emerson's followed his advice carefully, including the suggestion to keep a journal. Eventually, the younger writer moved in. Emerson called him "my Henry." In this sense, too, Thoreau joined a larger fray, because Emerson supported several Transcendentalists with his relative wealth (not truly self-reliant!). And Emerson in turn depended on him: Thoreau worked as Emerson's children's tutor, house repairman, and personal editorial assistant. His mother did his cooking and washing. Thoreau also spent quite a bit of his time in supposed isolation on Walden Pond in convivial social gatherings that would make many a contemporary American jealous. Although the current idea of him is as a man isolated in the wilderness, he was as highly networked as anyone could be at that time. Even during the writing of *Walden*,

Thoreau went into town frequently, conversing and dining with others and speaking about his work around the country, buoyed up in the popular eye by his connection to the Transcendentalists. Though he was understood to be a deacon of reclusiveness, in truth Thoreau received a startling number of personal guests during his stay, where he was within walking distance of the Emersons and also the railroad station. As he writes, in a less-often-quoted paragraph of *Walden*: "When I return to my house I find that visitors have been there and left their cards, either a bunch of flowers, or a wreath of evergreen, or a name in pencil on a yellow walnut leaf . . ." Or, as he writes of his Walden house, "I had three chairs in my house; one for solitude, two for friendship, three for society."

When Emerson writes, "Society everywhere is in conspiracy against the manhood of every one of its members," he himself is free neither of society's assistance nor of its constraints. In the summer of 1836, for instance, he started to meet his friends and colleagues in the Boston area in a clique ultimately called the Transcendental Club. He benefited greatly from the support of his community and was enmeshed with other thinkers of his age. While their mission was to enshrine the value of the individual, this didn't stop the group—which included his close friend the journalist and feminist Margaret Fuller—from being highly dependent on one another. At their regular gatherings, the squad conversed over a given topic—"American Genius" one week—influencing and supporting one another's thoughts. For one meeting, Emerson brought the club to his home, nicknamed "Bush," in the hopes of appealing to new members—he asked Fuller to "mould" the minds of the dinner guests. *The Dial*, the publication that members of the Transcendental Club created and contributed to until 1844, was also a testament to their interwoven lives and minds. Emerson was so connected to his circle of friends, in fact, that he begged Thoreau to move in with him; his second wife, Lidian; and his four children. He crowed that doing chores with Thoreau would make him "suddenly well and strong . . . [Thoreau is] as full of buds of promise as a young apple tree." Ultimately, they lived separated by a creek, which in autumn was filled with brackish plants, and brown branches, drying red berries and crows. Thoreau wrote *Walden*, published in 1854, on a plain, scarred writing table, distinguished by tiny finials and symmetrical shapes. Emerson wrote on a similar desk.

Like many who insist that they are self-reliant, Emerson could live as he did because of his relative wealth. He secretly depended on the fortune of his first wife, Ellen, for much of the funding that allowed him to work on his individualistic life of the mind. After Ellen's death, he was able to support his family and, on occasion, his friends after suing her people, the Tuckers, for his share of her large estate. He inherited $11,600 in 1834 (roughly $300,000 today) and another $11,674.49 in 1837, a total of roughly $600,000 today. He also received an income of $1,200 a year. Like most men of his time, he depended on his spouse to keep his home. Emerson's second wife, Lidian, is described as a "meticulous anxious housekeeper" who addressed her husband's fear of cold with new airtight stoves in every room of the house they shared and was so concerned about his comfort that even in her dreams she cleaned. In late February 1839, Lidian got the family house and "all of her possessions in absolute perfect order" while in the final days of pregnancy. She even cleaned the barn. By the next morning she had given birth.

To my guide, Howell, Emerson is "there in the people who pretend they are all temporarily embarrassed millionaires," meaning Emerson's version of individualism is also present when people who are down on their luck pretend they are merely having a small setback. In his book, Howell writes that he believes Emerson considered "the dependent" to be marked by "hopeless mediocrity," and that "entrepreneurship is superior to collective action." Howell says he himself is "against self-reliance," because, as he writes, "the ideology of self-reliance" can be understood as putting personal ambition and gain above society and the republic.

Emerson writes sentences like "Discontent is the want of self-reliance: it is infirmity of will." "The discontented" of Emerson's time included indentured servants (including Emerson's own), indigenous people, and women; that their unhappiness was simply the result of an absence of personal will is a clear fallacy.

Over the years, a veritable parade of intellectuals has tried to make sense of what is behind these writers' man-alone-'splaining. Some argue that Emerson was the architect of something particularly insidious: as one premier twentieth-century literary critic observes, Emerson is the mastermind of America's "imperial self," a writer who destroyed "shared communal hope," feeding the vituperative ideology of the Singular.

He didn't, after all, entirely join the abolitionists, although current scholars have complicated the image of Emerson as uninterested in the antislavery movement. As scholar James Read writes in his essay "The Limits of Self-Reliance": instead, Emerson "admired the self-reliance of individual fugitive slaves . . ." most of all and was most focused on a kind of mental independence of, in Emerson's words, liberating "imprisoned spirits, imprisoned thoughts, far back in the brain of man." As Read writes, "Emerson's philosophy of Self-Reliance provided less clear guidance for fighting slavery." In the last decade, the critic Kathryn Schulz also decried Thoreau as a kind of pinched proto–Ronald Reagan, an ungenerous soul who was "fanatical about individualism . . . convinced that other people lead pathetic lives yet categorically opposed to helping them."

Critics have also found him guilty of promulgating the corporate individualism that mirrored the political culture of the 1980s. It was fun to find the critic Leo Marx writing scathingly of Emerson in the *New York Review of Books* over forty years ago that his "tone anticipates that adopted by Reaganites in castigating lazy 'welfare mothers.'" Emerson has also been cast as an unwavering believer in minimum government. In Christopher Newfield's *The Emerson Effect*, the scholar questions him on different grounds, accusing Emerson of promoting "submissive individualism" whereby self-determination comes about after a person—well, a man—bows down to his societal superiors: individualism that was also a kind of conformity. For instance, in one passage Emerson sounds the graceless notes of an arrogant entitled bro, urging his readers to have "the nonchalance of boys who are sure of a dinner." He also could also be animated by even baser impulses, as when he delivered a backbiting eulogy at Thoreau's funeral in 1862. "He seemed born for greatness," Emerson remarked. "I cannot help counting it a fault in him that he had no ambition." Emerson went on to quip that Thoreau, his mentee and deepest ally, was merely "the captain of a huckleberry party." In his telling, Thoreau was closer to a pathetic underachiever, strangely content picking luscious berries by a stream.

═══

When I first read the works of Thoreau and Emerson in high school, I remember trying to force myself to love these two Great Men™ of letters as my teachers professed to. I had grown up believing in my own version of the self-made myth. Donning the armor of individualism meant reading as many books as humanly possible. I thought I'd at least *feel* shielded against the onslaught of a society that discouraged the success of so many girls. I had *Jane Eyre* as a counterpoint, but I also had grown up on a 1970s and '80s diet of rerun sitcoms, with their "lady drivers" and "jiggly" "girls" who were really thirty-year-old women. When as a young person I forced myself to embrace Emerson and company, it was also related to how I and other girls of the period were taught by the era's independent cinema (think of the silent men in fedoras in *Stranger Than Paradise*) to love indifferent but ostensibly poetic boys and men who despite their literariness barely spoke.

To be fair, the version then of these men wasn't just the Transcendentalist-fanboy rendition. I was taught in various classes that they were also avatars of a kind of individualism that was all about necessary and healthy rebellion against oppressive and conformist hierarchies, like that of the Catholic Church. This kind of independence meant you were on your own yet you were also cool with civic ties and you still gave a damn about others.

In this reading and related ones, Emerson et al. were also supposedly espousing renegade freethinking, the equivalent of a brave man standing up to a town full of normcore minds. In other words, Emerson's idea of self-sufficiency was not just that of the contemptuous woodsman or the macho coder or a guy who thinks taxes are for losers, but the self-reliance of the freed colonial subject, the independent-minded, the nonderivative, and the outcast.

Some examples of this positive reading include that of the eminent Emerson biographer Robert Richardson, who writes that his subject's work "is not a blueprint for selfishness and withdrawal; it is not anti-community" but "recommends self-reliance as a starting point" rather than "a goal." For instance, the scholar Jedediah Britton-Purdy writes in 2015 of Thoreau's "'doubleness,' a sense of observing his own life from alongside": that through this technique Thoreau could be taken to be arguing *against* what he appears to argue for and thus that "Thoreau's [communitarian rather than individualist] idea that injustice in the community touches everyone."

Unlike the businessmen who wrote crowing memoirs in the 1980s bragging of their self-made wealth, the writer of *Walden* is and was often also explained as an exponent of being above the fray of quotidian existence. Childless, unmarried, existing only for his relationship to Nature and the Word, Thoreau and, to a lesser extent, the more familial Emerson were pitched as paragons.

Even as a teenager, I had a sense that *Walden* and Emerson's essay "Self-Reliance" were selling me something that could hurt me as a young woman of modest means who needed her peer group and interdependence and solidarity to succeed. There was a chance that these texts would cancel me out if I truly absorbed their message, but I forced myself to let them influence me. Teachers and professors thought of him as magical, dreamy, democratic Emerson, the guy with Big Male Heavy Metaphorical Energy (according to the lionization of him by the intellectual Irving Howe and others). In graduate school, in seminars, passionate and frustrated doctoral students who were teaching multiple classes at community colleges for what amounted to minimum wage were asked to believe Emerson when he wrote misty things like: "With consistency a great soul has simply nothing to do. He may as well concern himself with his shadow on the wall."

As I read this, I was working all the time—teaching, tutoring, reporting, pitching, writing, fact-checking. I worked odd jobs like taking an older women's business correspondence and packages to the post office for her and cleaning up the gibberish corruptions that populated digital documents. I was often ashamed as I thought of my provisional and economically unstable career at the time as embarrassing, an example of the way I wasn't a functional part of society. (I was critical of myself for this but not of others.) This was part of why Emerson lured me in with his reverie-laden prose, feeding me ideas that would curdle inside of me, culminating in the belief that if I ever stopped working or achieving, I would disappear.

———

My time with Howell concluded with his giving to me a concept of autonomy and its opposite that was most unlike the faith of the famous men we were considering: he called it the "arts of dependence."

Howell deployed the term to describe young women in nineteenth-century America who weren't, like Emerson and Thoreau, even allowed to paint themselves as self-reliant. Instead, Howell argues, young women were likelier to do work that was considered then to be minor or imitative, like the embroideries created by so many now-anonymous students at Mrs. Rowson's Academy for Young Ladies in Boston. The students at Susanna Rowson's famed school, founded in

1797, might embroider, say, a picture of a woman mourning. They would use delicately colored fine silk thread—the results rivaled those of painted portraits. Such works, Howell writes, are as important as the "independent art" of the Great Men of American letters, though most of these female embroiderers labored in anonymity. The young women created their botanical or galactic stitchery from other people's images, and they did so in a klatch, not making any claims for uniqueness. These works sustained their creators and either equaled or exceeded what could be created by a single artist.

I suggested to Howell that the concept of arts of dependence could be broadened, spanning not just the production of craftspeople of the past, who made things together without airs, but also to something I thought of as an "art of interdependence." That phrase could describe the skill it takes to depend on others or even to receive government assistance—the resourcefulness and craft it takes to collaborate with others and also to take a position of dependence that has as its requisite a heightened engagement with society. It takes aplomb, for instance, to learn how to build a house and farm on government-provided land; to feed a family of five on minuscule monthly Supplemental Nutrition Assistance Program (SNAP) benefits; or to navigate crosswalks while seated in a wheelchair, even if the curbs are designed with regulation cuts.

If we could adopt and internalize my revision of Howell's phrase and really think of interdependence as an art, we could recognize that we are connected and mutually dependent not just on our families, but also on all of the institutions that support us, from schools to hospitals, and understand that accepting aid is a kind of grace.

Howell, for one, knew how important interdependence is, both personally and intellectually. Part of it was that he had two very young children, and during the week they were in day care. Like many Americans, including me, he and his family relied on day-care workers, who were often paid less than dog walkers. "I am a total dependent," he said. "As someone with kids, if I didn't have everyone else helping me, my life would simply not work." The "everyone else" comprised the day-workers, his wife, their parents, and their colleagues and bosses who tolerated or even encouraged taking sick days.

The sun had gone down though, and he had to leave to have dinner with his children. As I stood in front of Emerson's home, where there are faint marigold-yellow mini-LED lights in every window casting an eerie glow, I recalled the essayist Geoff Dyer's book *Out of Sheer Rage*, a personal favorite, and the scene where Dyer visited the home of his book's subject, the novelist D. H. Lawrence: "We stood silently. I knew this moment well from previous literary pilgrimages: you look and look and try to summon up feelings that don't exist. You try saying a mantra to yourself . . ." I felt that way at Emerson's house too. But what I got from my visit was a reminder that he was one of a network of thinkers who helped set in motion a train of thought and ideology. From literary champions to pulpy populists they helped forge the self-punishing individualism we have today, which has become an all-consuming movement.

3

Little House of Propaganda

A FAMILY TRAVELED in a covered wagon, a mother, a father, and three young children. After miles and miles were logged, the fivesome decided to settle down in Missouri. Not long after, the father pleaded with his wife to let them leave that state and they proceeded on another leg of a trek, envisioning a future farther and farther west. They got to Wisconsin and then Kansas, part of the conflux of American pioneers. "We can get a hundred and sixty acres out West, just by living on it, and the land's as good as this is, or better," the father told the mother. "If Uncle Sam's willing to give us a farm in place of the one he drove us off of, in Indian Territory, I say let's take it. The hunting's good in the West, a man can get all the meat he wants."

In 1874, the family arrived at its final destination, Minnesota's Walnut Grove. "Pa" Charles lands a job. They build a log cabin. Pa spends his off hours farming on his own. At Christmas, Pa and Ma have to take additional jobs so they can afford gifts for their three girls. One of Charles's jobs—as a powder monkey at a quarry—is unusually risky, but he does it for the added income. The family's hardships don't stop there. They also encounter terrible weather, as when a hailstorm destroys the crops in the area. Sometimes the unpleasantness is interrupted by pleasures: a barn dance, the children gathering snow and covering it in maple syrup to create makeshift ices.

All of these scenes are drawn from Laura Ingalls Wilder's *Little House on the Prairie*, just one of that autobiographical series of eight books, all but one published from 1932 to 1943 (the last one was published posthumously). To date, the series has sold over sixty million copies. If Emerson and Thoreau helped establish the poetic and intellectual version of self-sufficiency, Wilder espoused a mass cultural, child-friendly version of it. The series wound up in millions of living rooms and in the hands of young girls, forming their values. We might call the result young-adult bootstrapping.

The books were undeniably appealing, trading in effective sentimentality. Wilder's novel *The Long Winter*, for instance, about how her family survived a savage winter in the Dakota Territory, making sticks from hay to serve as coal and grinding their own wheat in a coffee mill for bread, is catnip for an intrepid twenty-first-century child or tween.

Wilder's *Little House* is girlish tessera in the mosaic that is the pioneer-Western-self-creation fantasy. It is a do-it-yourself-in-the-territories narrative so enticing that *Little House* stayed with many of its young readers into adulthood. As one such person who grew up on the edge of the Great Plains writes, they were a source of "poisonous cultural worship of bootstraps . . . that I absolutely treasured." The books could be seen as underlying both an American pressure to make it on one's gumption alone and the accordant shame when a person could not do so. What *Little House on the Prairie* indicates and suggests more broadly is that even we denizens of the twentieth and twenty-first centuries—who are not able-bodied white men willing to take land from indigenous populations—should follow Wilder's mythologized parents' example and make our own reality with our own hands or be deemed weak and failed.

Wilder's own biography hews closely to the lives depicted in her series. She was born in 1867 in a log cabin in rural Wisconsin. When she was twelve, her family became homesteaders, moving frequently. She and her siblings were partially self-taught as a result. The family always needed money, which was why at only fifteen she started working full-time as a teacher. As in her books, she married an older family friend, Almanzo Wilder, at eighteen. The couple had two children—one died. The Wilders lived a peripatetic life, traveling from place to place, stopping finally in the Ozarks, where they ran a farm and did all the work.

Laura's literary legacy, though, was sealed by her daughter, Rose Wilder Lane. Lane pushed her mother to publish her autobiography. A reporter and a rabid libertarian to boot, Lane worked with Wilder on the manuscripts and mentioned in her diaries that she heavily edited them. For all of Lane's aggressive ambition, there was no great success at first, because Wilder's initial autobiography was rejected. In 1932, the first book of her series was published and became an international megahit. Laura Ingalls Wilder would continue to use her fame—her books, her lectures—as a bullhorn for the bootstrapping paragon, until she died in 1957 at ninety years old.

In her books as well as her public life, Wilder loudly trumpeted the self-reliant ideal. She found a corollary in the political culture defined by President Herbert Hoover and others. In 1928, Hoover used the phrase "rugged individualism." In a campaign speech that year, he deployed it and then offered himself as its poster boy, rather than espousing the "European philosophy" of "state socialism" that he implied belonged to his rivals. Hoover declaimed: "The American pioneer is the epic expression of that individualism. . . . That spirit never need die for lack of something for it to achieve." (The Herbert Hoover Presidential Library and Museum houses Wilder's daughter Rose Wilder Lane's papers and testaments to what its website calls her "extraordinary life . . . formulating and promoting libertarian ideas," which included her positive biography of Hoover.)

Hoover's rabidly bootstrapping stance was refuted by Franklin Delano Roosevelt's positive attitude toward social programs, which was part of why Wilder stood in opposition to that president. Libertarians *avant la lettre*, Wilder and Lane hated FDR and saw the New Deal as weakening the citizenry. "We have a dictator," Lane writes of Roosevelt in her journal, also saying she'd like to "kill that traitor." She must have been hopping mad when Roosevelt countered his predecessor Hoover's rabidly individualistic rhetoric in 1935 with oratory encouraging Americans to "unlearn" that "superstition" that "the American spirit of individualism—all alone and unaided by the cooperative efforts of government—could withstand and repel every form of economic disarrangement and crisis."

The Depression ultimately catalyzed radical social solutions and changes, among them the cooperative-minded New Deal. For writers and artists who, unlike Wilder, were desperately poor, FDR created the Federal Writers' Project, which ran for eight years starting in the middle of the Depression and was part of the Works Progress Administration. It supported 6,600 writers, editors, and researchers who created not only journalism but also oral histories, including first-person recollections of formerly enslaved people and stories of Depression-era Hoovervilles, or shantytowns, created by the previous president's draconian housing and economic policies.

Wilder, meanwhile, saw her books as an argument against these and other governmental assistance programs that arose at that time. She publicly commented that government was too intrusive, that taxes were part of "the swift and violent death" of democracy.

═══

When my daughter was seven, I reentered Wilder's world. I would stay up watching the TV show based on the books, which I read to her out loud. Our protagonist, Laura, was indeed spunky (as they used to say of girls), and her family's adventures—surviving when it was too cold and stormy to see even their neighbors—engrossed us.

My daughter loved *Little House on the Prairie* so much that she decided to dress up for Halloween as Wilder, her idol. She donned a crisp poplin dress, plaited her hair, and gently wheedled me into wearing a white bonnet and purchasing a corncob pipe and suspenders for my husband, an unlikely Pa, especially in his 1940s fedora. Rugged individualism-as-costume party: my daughter liked the idea of independence that she found in the books and the TV series; it appealed to a child with a sharp will who nevertheless still very much needed her parents. She had been a lifelong apartment dweller who believed city parks were nature, but she had fallen in love with the Wilder-verse in all its guises. As we watched the show together, I couldn't help but recognize that though they were seemingly harmless, these novels and that hit show were a crucial example of the way the bootstrapping ideology wormed its way into the minds of children. The books and show gave her an image of a family thrown back on their own resources. She—and her generation's children—might long for this intensely. Part of the appeal was the insularity of the books, and their cozy, *hygge*, vibe—like a house themselves, they were built around a single family and appeared on the surface like a log cabin, airtight and humble and reassuringly old-fashioned.

But I knew there was a dark undercurrent to these books and that they rested on some false premises. When the Ingalls family, for example, moved to Kansas (before Minnesota) to homestead, it wasn't just the family's pluck that served them—that they could get land there at all was due to white homesteading. By 1877, most of the Kansa or Kaw Indians had been pushed from their land to the Indian Territory. As the Oklahoma Historical Society puts it, "The end of the Civil War allowed another surge of Anglo-American settlement into the West, and again tribal nations were pressured onto reservations in the Indian Territory," laying the groundwork for the Ingalls family and those like them.

Laura Ingalls and her sisters and her mother can—and must—make their own dresses and start their own fires to survive, at the same time never acknowledging the ways in which white families had access to this "independence" only because they took indigenous land for free. To seal in the denial of the fact that the land they were living on most likely had been tribal, the most famous of the *Little House on the Prairie* series of books contains variations on the slur "the only good Indian is a dead Indian" not once but three times.

In addition, for the women themselves, Plains and pioneer existence often felt less, well, free than Wilder liked to imagine. Feminist author Carol Tavris puts it like this in her book *The Mismeasure of Woman*: "Was the American frontier 'conquered' by single scouts, brave men 'taming' the wilderness and founding a culture based on self-reliance? The mythic vision excludes the women who struggled to establish homes, survive childbirth, care for families, and contribute with men to the community."

Curious about what life on the Plains was really like for women, when my daughter went to sleep I read a version of pioneer life that showed me a far more authentic picture, the 1929 autobiographical novel *Daughter of Earth*. Its author Agnes Smedley's family lived in a log cabin on the Plains and in a mining camp in Colorado. The family barely survived hunger and a cyclone. In *Daughter of Earth*, Smedley, born in Missouri in 1892—twenty-five years after Laura Ingalls Wilder—wrote of the protagonist of her novel that was inspired by Smedley's own life and how her fictional stand-in, Marie Rogers, and Marie's mother and sister washed clothes to make ends meet. Their house was "one mass of steaming sheets, underwear and shirts, and to get from one room to the other we had to crawl on the floor."

The mother in Smedley's novel cobbles together dinners of potatoes and flour-y water, paid for with the money she earned leaving home at seven in the morning for her job as a washerwoman. Smedley writes that her hands were "big-veined and almost black from heavy work." (Mother drops dead at thirty-eight.) Pa is also violent, abusive, itinerant, useless. At one point, they live in a town that is more or less owned by a monopolistic mining company and that circulates around a dominant company store, "the one store from which we bought food and clothing . . . no others were permitted to exist."

At the age of thirteen, Smedley and her protagonist both worked in a series of after-school jobs she "despised," including as a teenage sleep-in boardinghouse maid, "fired for drinking her employer's milk," and a tobacco stripper. Like many working-poor women of that time, the women in *Daughter of Earth* labor like oxen, swallowing their humiliations (I think of my own grandmother as a teenager at a similar time, sticking feathers on hats in a New York factory).

The only woman with any self-regard is the protagonist's aunt, a sex worker whose exploitation is clear-cut and whose profits are her own. In fact, Smedley made the argument over and over that the life of a prostitute in the nineteenth and early twentieth centuries, especially in these Western towns, was often preferable to that of a working-class wife. The sex worker, if she didn't work for a violent pimp, according to Smedley, was less likely than, say, a homemaker or a washerwoman of that time to be beaten by her husband. Smedley's "poor farming woman" story has a rough honesty that is unlike Wilder's airbrushed-with-gingham reminiscence. It's a portrait of a time when women in few states had the right to vote—in Smedley's novel they have just received that right in Colorado, but the mother's husband doesn't permit her. The only way out for either the mother or the aunt or for Smedley herself is—no spoilers here!—a mass solidarity movement.

And thus Smedley, once she escaped her pioneering childhood, became a well-known and then an infamous Communist sympathizer.

═══

I was hoping my daughter would see the faults of Wilder's books and the TV show based on them and, widening out, in the rugged individualism racket as a whole. I also wanted her to understand the problem of racism toward indigenous people in the books and the show. I mused to my daughter that celebrating the straightforward plunder of land might not be a great look. (Also, what do we really think of a story where no one ever relaxes?)

At the same time, I also tried to impart to her the stray good lessons in the books—like persevering in school. I wanted my daughter to understand a different notion of independence than the one propagated by *Little House*. I wanted my daughter to understand that *dependence* and *interdependence* should be included in all of our understandings of our own autonomy. It was an understanding I had partially thanks to contemporary activists like Sunaura Taylor. At a disability-rights march a number of years ago, Taylor was having trouble eating without a table (she has limited use of her hands), and a woman told her: "You can ask for help here. This isn't a place where we value independence." Taylor and others like her want us to understand that able-bodied-ness permits so many of us to maintain the illusion of independence, but that it's a transient state.

Even if we think we are independent, now or at some point in the past or future we were or will be utterly reliant on others.

This was true in the real life of the woman who helped create the autonomous pioneer archetype. Wilder and her father, Charles Ingalls, as well as men like him, depended on others. In Wilder's family's case, they also leaned on the government, for instance through the Homestead Act of 1862, the federal-assistance program that offered, mostly to white American men and their families, 160 acres each of Western land, appropriated from Native Americans. Indeed, the Homestead Act could, in one light, be understood as the ultimate handout. In what can be a pattern for those who claimed utter self-sufficiency, Wilder occluded her family's dependence, not presenting it to her readers. Whether this omission was conscious and sneaky or such a universally accepted form of falsification that Wilder didn't need to actively suppress anything is not clear.

But despite the land's ambiguous ownership, she and her family were beneficiaries of the Homestead Act. It is considered by some experts to have arisen out of earlier grassroots efforts, or like the mass activism of farmers in upstate New York in the 1840s wherein fifty thousand rural tenants facing eviction suddenly reconsidered—and wildly protested—privately held land. As one scholar told me, both the Homestead Act and the earlier Anti-Rent War rested on "a notion that land is not created by human beings and they had a right to it." Wilder ignored this history, crediting her family's success to much-ballyhooed "resourcefulness."

Her family was able to put down stakes and prosper out West on their land as a result of governmental intervention and the inborn privileges that allowed her family to be recipients of that largesse. One of those automatic privileges was that she was white and the homesteads were given almost entirely to whites. And even though after 1865 and emancipation, land *could have* been one of the main reparations that Blacks received, they were instead prevented from claiming the places where they had long worked as slaves. (This exclusion is, in fact, an origin point of Black landlessness, lasting to this very day.) The promise of land for Caucasians attracted the Ingalls family to Kansas in 1869, to a 4.8 million–acre tract on the Osage Diminished Reserve. Ultimately, the family would make a homestead claim in De Smet, in the Dakota Territory. The family was one of the recipients of "one of the largest federal handouts in American history," as author Caroline Fraser writes in her vivid biography *Prairie Fires: The American Dreams of Laura Ingalls Wilder*. Thanks to the Homestead Act—a benefits initiative by another name—10 percent of America's land was given out as homesteads to up to a quarter of America's population. A total of 270 million acres was distributed to 400,000 families, or about 1.6 million Americans. It was a "massive transfer of wealth, one of the largest in American history, to everyday people to provide for their families," as Mike Konczal, a Roosevelt Institute economist, writes.

Given that her family were in a sense sustained by the American government in terms of their property, the *Little House* books and Wilder's personal history could have been reframed as the story of our country's care for its citizens. After all, Pa could be referring to either the Homestead Act or the add-on Timber Culture Act of 1873 when, in *By the Shores of Silver Lake*, he says, "This country's going to be covered with trees . . . Don't forget that Uncle Sam's tending to that." But it was not to be.

It must be said that Wilder and her family were not anomalous for seeing their achievement on the Plains as evidence of their individual powers rather than government assistance. After all, the giveaway of lands—and the myth that these pioneers suffered or flourished due to their mettle alone—was part of what inspired Wilder to create her series of novels in the first place. (On a pettier note, Charles Ingalls was dependent and also interdependent, because evidence suggests he was not such a great farmer and thus leaned on his neighbors for help far more than Wilder tended to admit in her books.)

It was also easier for Wilder to tell an individualistic story in the twentieth century, long after the Homestead Act's generosity had been absorbed, sucked up by speculators claiming some of the best public free land: big capitalists went to the trough and feasted on what one early-twentieth-century historian, Vernon Parrington, called "the great barbeque" of these land parcels.

═══

These events had long been consigned to the past when, in the 1960s, Wilder's daughter Rose set out to sell a different kind of property: the television rights to her mother's novels. While peddling the books, she insisted on fealty to the text. Upon her death in 1968, she bequeathed the rights to a friend, Roger Lea MacBride, and he ultimately developed them for TV. As in the books, the adorable family on the Plains was the beneficiary of the government's munificence through the Homestead Act and, like the books, the hugely popular show—NBC's highest-rated scripted program until 1982—cast the family's success as due to the triumph of their individual wills. Pa was played by Michael Landon, who had changed his name from the Russian Jewish Eugene Orowitz (Landon's own self-made story). The Pa role was a large part of what put Landon on the cover of *TV Guide* twenty-two times.

I first watched the show in the New York City of the 1980s, a lonely only child of the Reagan period. To my young eyes, Reagan and Michael Landon had a similar suntanned affability, Pepsodent smiles, and saccharine kindness, discordantly crossed with macho fervor. I'd see those qualities sitting right up close to the flickering, squat square of our family's Panasonic TV.

The nineteenth-century West of the television show *Little House*—actually a set in Southern California—evoked Reagan in another sense. Episodes of the show echo the punitive morality of Reagan's harshest speeches. The show's hobos and drinkers couldn't cope and were punished with early deaths, while Pa and Ma continued to prosper. I didn't understand it as a child, but both the series and many of Reagan's speeches presented the fiction that help can come from no one but ourselves.

Forty years later, I am not surprised that *Little House,* the book and the television show and the ideology, continued on for all of these decades and was even slated to be rebooted. It also made sense that its vision had pollinated other entertainment products as well. When my daughter outgrew *Little House,* she became captivated by a program that seemed to emerge from the same belief system, and this time so was I. It was the survival reality show *Alone* on the cable channel History. Contestants are dropped on their own in Mongolia or the Canadian Arctic and must find a way to sustain themselves, catching their own food, building their own lean-to dwellings, filming themselves as they starve or slowly go mad. The contestants drink squirrels' blood to survive and film themselves doing so, all to win $500,000. This show and others of its ilk are like an unwitting parody of *Walden* or Wilder on Adderall. We watched this fare so much that my daughter knew how to make a homemade teepee out of branches when we went to the park. *Alone* mines the individualism myth of the American West the way *Little House* does. It is a story whose distortions—toxic in myriad ways—continue to hold sway and do damage. There were studies and theories, for example, that connected pioneer narratives to counties with poor public health practices during COVID. One study published in March 2022 in the *Review of Economics and Statistics* found that the more individualistic counties in the United States, many of them formerly frontier areas, engaged less in social distancing and were also less willing to receive COVID-19 vaccines. In another paper entitled "Rugged Individualism and Collective (In)action During the COVID-19 Pandemic," from 2020, by a different set of researchers, economist Samuel Bazzi among them, also found that counties "with greater total frontier experience (TFE) during the era of westward expansion" had a "weak collective response to public health risks." The researchers concluded, "Frontier culture hampered the response to the COVID-19 pandemic." This data made it even harder to accept the romanticization of the rugged individualism of the past.

In addition to its continued effect on television shows and public health, the drumbeat of individualist pioneer propaganda was also a centerpiece of some key political speeches. These speeches might be broadcast on television, like one delivered by then-President Donald Trump at the 2020 Republican National Convention. "We are a nation of pilgrims, pioneers, adventurers, explorers, and trailblazers who refused to be tied down, held back, or reined in," said Trump. "Americans built their beautiful homesteads on the open range."

"Those homesteads were a gift from the government!" I said to the television screen, as my daughter sat on the couch beside me. She was, thankfully, exhausted by the celebration of these pioneers, so she immediately interjected, "What about all the Americans who didn't get land?"

4

The Horatio Alger Lie

DICK IS A teenage bootblack who polishes the shoes of far older, richer men and at night sleeps on New York City's pavement, sometimes under wagons whose drivers are also asleep. We are told he has an open face and the character of a perpetual innocent despite his hard, mean surroundings. Ultimately, Dick obtains a well-paid job as an office clerk and goes by the more mature and respectable name Richard Hunter, finally cutting himself off from "the old vagabond life."

Dick/Richard Hunter is the protagonist of *Ragged Dick,* the most famous novel by the nineteenth-century author Horatio Alger Jr., whose oeuvre—fictions of young men rising rapidly in the ranks— gave rise to the idiom the Horatio Alger story. These penniless boys and men who make it up the ladder of American commerce do so through pluck and hard work. If we abstract it further, a Horatio Alger story is an idea of American life that starts with indigence and ends in affluence.

More than a hundred years after Alger's death, the phrase is still such well-known shorthand for the self-made myth that the documentary filmmaker Michael Moore said that Americans are on "a Horatio Alger fantasy drug." The reality TV personality and cosmetics company owner Kylie Jenner, half sister of Kim Kardashian, absurdly, was profiled in the *Chicago Tribune* under a headline that partially read "A Horatio Alger Story for Our Time?" (the question mark clearly doing too much work here). And Alger's name has been used to describe Michael Bloomberg, New York's billionaire former mayor and presidential hopeful, who was born quite middle class.

It originated with the tales of the immensely popular Alger himself, who wrote dozens of books with that rags-to-riches plot. After his fourth young adult book, *Ragged Dick,* he wrote roughly one hundred other novels, becoming one of the most influential writers of his generation. By 1910, his young adult novels were selling at a rate of one million per year. And throughout the rest of the twentieth century, his name was applied widely, and people still more or less knew the author who was its origin point. (An example from the infamous 1987 memoir *Trump: The Art of the Deal* that the former president—by way of his ghostwriter, Tony Schwartz—writes of his father, Fred Trump, a housing developer in Queens: "His story is classic Horatio Alger.")

Today, while the Alger conceit continues at full force, many of his books are out of print: I found only a few on the high shelf of a private library, first editions with inky black-and-white illustrations of boys with lithe physiques. In these novels, the central male characters start at the bottom of the pyramid as stock boys, bootblacks, and immigrant buskers. The books have titles like *Tony the Tramp*, a young, lost boy who becomes a millionaire by the end and gets his rights to his English estate back, or *Paul the Peddler*, who sells men's neckties and then steers his way to solvency. Yet almost all of these books are, if read closely, not Horatio Alger stories at all. Rather, they run on the idea that every self-made young man in fact has a hidden dependency on a more privileged person, usually an older man. In *Ragged Dick*, the eponymous hero befriends respectable older men—Mr. Greyson and Mr. Whitney—who cross his path by sheer luck, and they help him make it off the streets. He ultimately makes it out of poverty permanently due to special treatment by James Rockwell, a rich patron and industrialist.

In other words, in Alger's books the poor young boys *do not* make it entirely on their own. As Alger scholar Carol Nackenoff told me, Alger's boys "escape precarious financial circumstances" because they are "taken notice of by someone able to give them a chance," usually an older rich gentleman. They wind up having to depend on an older rich person, because there is no other systemic support. Embedded in Alger's bootstrapping myth appears to be the use of problematically intergenerational attraction to boost one's social status. This was perhaps part of why the novels were set in an earlier, less industrialized time that would allow boys to run into men who might rescue them.

Alger himself was not so far from the Horatio Alger story biographically. Though a Harvard Divinity School graduate, he got famous and rich on his own early due to his own penchant for publishing book after book—literary overproduction, I suppose, could be considered its own version of bootstrapping.

But Alger's life didn't follow a Horatio Alger story in a crucial way: Alger had a very public fall from grace when he was young and discovered to be a pedophile. Alger's pedophilia, in fact, was the buried truth of his life and the tainted centerpiece of his unwritten biography, not the one found in the well-known Ralph Gardner biography *Horatio Alger: Or, the American Hero Era*.

This secret was exposed before Alger published his more popular books, back when he was a parson at a church and discovered to have indulged in "unnatural acts" with two young boys. A member of his congregation, Solomon Freeman, writes in an 1866 account that when a young boy "called at [Alger's] room to leave a book . . . [Alger] bolted his door" and then "committed this unnatural crime." One of his victims was a thirteen-year-old who told his parents the new minister had molested him. The other was fifteen. Alger was accused of "the abominable and revolting crime of gross familiarity with boys," in the words of a parish report to church hierarchy. Alger admitted at the time that he was "imprudent." He was, according to records, "run out of town by a howling mob." But it was a touch less dramatic than that. His father, also a minister, told officials of the church that his son would never again seek a position in the ministry if they allowed him to quietly leave.

Alger resigned from his Congregationalist church, slinking out without a mark on his record, likely due to his own relatively privileged status as the minister's son. Alger then spent the rest of his life creating fictions about ephebes and young men, and older fellows like Mr. Whitney and Mr. Greyson helping boys get off the street.

The more I researched Alger, the more I believed "bootstrapping" was a story line emerging out of the ashes of Alger's own personal shame for his pedophilia and the trauma he inflicted on others. His life story morphed from his longing for young boys into the ultimate story of young boys' exerting mastery over the adults and the world around them, perhaps his own identification with his powerless victims.

As James Martel, a professor at San Francisco State University and the author of the paper "Horatio Alger and the Closeting of the Self-Made Man," put it to me, Alger's "obsession with self-reliance masked his repressed sexuality," as Alger "sublimated his desire for adolescent boys into a lucrative literary career writing tales for young male readers."

As Michael Moon observes in his 1987 essay about Alger, "'The Gentle Boy from the Dangerous Classes': Pederasty, Domesticity and Capitalism in Horatio Alger," in "each book, a boy is 'saved from ruin,' from possibly becoming a criminal or a derelict, by being fostered as a candidate for recruitment into the petty bourgeoisie." The American public has forgotten the tawdry and illegal Alger backstory if they ever knew it. The outré parts had been whitewashed.

After he was run out of the ministry in Brewster, Massachusetts, in 1866 (he was brought up in Chelsea, near Boston), the boyish author, five feet two at full height, began to churn out novel after novel, from Westerns to what now might be called young adult literature. He found some of his material as a result of being fixated on the thousands of neglected children in the streets of New York City— the urchins—an activist cause célèbre at the time. He spent his free time observing these homeless kids—one historical observer put their number in New York City in the 1870s at tens of thousands—and took the seeds of his most successful books from New York City's lost boys themselves, who were selling newspapers and hanging out on the docks. Alger was reputed to gather the boys to him like the Pied Piper, giving them money and candy. In his rooming house on St. Marks Place, Alger held a proto-meet-up for them. We have no idea how he behaved with them in private or with the two street boys whom he eventually adopted and housed in his apartment: they had far less power to be heard than small-town churchgoing boys in Massachusetts with their nuclear families intact. Whatever the case, it's clear Alger found inspiration for his novels from these youths as well as the adult muckety-mucks he knew. His friends ranged from the waifs he tutored to the most famous literati of his day.

In addition, the economic contours of his historical moment shaped his plots. Alger found his literary career after the end of the Civil War, when most of the rich came from wealthy households: he was writing in the spirit of his time, which was suffused with a longing for social mobility. Finally, Alger found inspiration in the founding fathers of self-sufficiency, like Benjamin Franklin. Two of Alger's novels were based loosely on the life of Franklin. Franklin, in his autobiography, described how he rose up, creating a printing house by doing even the basest work entirely himself—collecting reams of paper, which he would push down busy city streets in a wheelbarrow. In *Poor Richard's Almanack,* Franklin extols hard work as the ingredient for societal buoyancy, as he quotes Proverbs 22:29: "Seest thou a Man diligent in his Calling, he shall stand before Kings."

These grander inspirations and Alger's years of popularity didn't help him retain his standing in publishing. Toward the end of his life, he wrote biographies of American presidents, but they were considered overly potboiler-ish.

Nevertheless, after his death in 1899 his reputation grew. The *Saturday Review* in 1946 compared soldiers coming home from World War II to Alger's characters, even when they benefited from one of the great social supports, the G.I. Bill: in "Horatio Alger fashion, so to speak, the veteran will be able to pull himself up by his educational bootstraps." Two decades or so later, Ronald Reagan intersected with the Horatio Alger story in more ways than one, winning the 1969 Horatio Alger Award when he was governor of California, his personal rags-to-riches story part of what was said to make him Algerian: "Reagan's father was an alcoholic who lost his job on Christmas Eve, at the onset of the Great Depression, and struggled thereafter to support his family . . ." (Reagan had, in fact, exaggerated the extent of his father's alcoholic desperation for bootstrapping effect.)

In all of these moments, the Horatio Alger story, like "pulling yourself up by your bootstraps," was synonymous with the story that hard work brings about success, rather than admitting to a more complex plot. It's a story that feeds into popular opinion of our day: according to a Pew study released in 2020, 53 percent of Republicans and Republican-leaning people say that "hard work" is the explanation behind wealth, despite the omnipresence of income inequality that tips the scales no matter how exhaustively one labors.

Today, the Horatio Alger Award and Association still exist, and the group has rewarded a motley crew from Dr. Norman Vincent Peale to Ronald G. Wanek, the founder and chairman of the board of Ashley Furniture Industries Inc. It is the largest furniture manufacturer in the world and also has a Horatio Alger–ian secret. In 2015, while Wanek was at the helm, the company racked up $1.76 million in fines for Occupational Safety and Health Administration (OSHA) violations, and within three years, at the company's Wisconsin factory alone, there were more than a thousand worker injuries, according to the OSHA site.

Meanwhile, as strange as Horatio Alger's own biography, novels, and the association may seem, even in 2022, the Alger myth lingers.

———

In the middle of the twentieth century, Horatio Alger had pungent competition. Her name was Ayn Rand. Rand, a novelist and impresario, openly loved personal and political selfishness. She promoted an unhindered free market, where the economy was a growth machine on autopilot.

While a slightly more systematic thinker than Alger, she was not a more rational one. Nevertheless, for decades, she has shaped the minds of masses of people. Her books *The Fountainhead* and *Atlas Shrugged* have sold roughly thirteen million copies in the United States alone. Polls have found *Atlas Shrugged* to be the most influential book for readers after the Bible, and readers in another survey said they considered *Atlas Shrugged* to be number one among the greatest novels of the twentieth century.

In these works and her public appearances, her message was as simple as a company logo and as consistent as binary code. As she puts it in her early journals, "If man started as a social animal—isn't all progress and civilization directed toward making him an individual?"

Like Alger's pulpy messaging, Rand's diagrammatic characters, plots, and prose are a key to her lasting power. Her coarse simplicity helps explain why teen boys and billionaires manage to focus on her books long enough to make them their religion.

Why should I consider Rand here, in the same breath as Alger? Because she helped define how certain moneyed classes think about their power and was one of the premier faces on the proverbial postage stamps of tech individualism. Her effect could be said to be Alger 2.0, and as with Alger, her favored vehicle for transmitting this American archetype was pulp fiction.

Her contemporary followers are legion and include the King-of-the-NASDAQ-by-way-of-Burning Man Travis Kalanick, cofounder of Uber, who had the cover of a Rand book as his Twitter avatar, as well as the founder of Snapchat, Evan Spiegel, and Twitter's Jack Dorsey. Lisa Duggan, in her account of Rand entitled *Mean Girl,* writes that the most "influential figure in the industry, after all, isn't Steve Jobs or Sheryl Sandberg, but rather Ayn Rand." Apple's cofounder Steve Wozniak even called Rand's *Atlas Shrugged* one of Jobs's "guides in life." Rand's version of self-made absolutism is particularly attractive to these people, because they tend to be more absolutist: no one may contradict their point of view. As Adrian Daub wrote scathingly of her allure, "Who is teaching them [bro-grammers] that when they press a button on their keyboard, millions, or even billions, of people can be affected, sometimes in terrifying ways? Is Ayn Rand their only guiding light?" All that Rand was missing was a black turtleneck and a clean interface.

One suggestion for why the tech masters of the universe who like things like 3D-printed hotels buy into Rand's rendition of bootstrapping is that she offers "an understanding of self-reliance that doesn't really stand up to scrutiny once you've had to, you know, actually self-rely," as Daub snipes at Rand in his book *What Tech Calls Thinking.* Daub believes that Silicon Valley's preoccupation with prodigy and young entrepreneurs—a veritable fountainhead of youth—also chimes with Rand's ideas.

It's not just powerful technologists who have for so long used Rand as a source, swaddling corporate existences in libertarian twaddle. Rand inspired Whole Foods Market's CEO John Mackey, who finds total freedom in organic tomatoes. The holistic free marketer floods his interviews with musings on his "enlightened self-interest" and argues against accessible health care for all, saying that "the best solution is to change the way people eat. . . . A bunch of drugs is not going to solve the problem." She has moved investor Mark Cuban, who named a boat after one of Rand's novels, and Lululemon's founder and former CEO, Chip Wilson, who publicly celebrated Ayn Rand's libertarianism along with his company's compression leggings. Rand is part of the banality of affluence, where providing profits to a tiny sliver of the population is an example of what she called "to gain and produce." Rand also seeped into the bootstrapping attitude of total freedom held by some who work in private equity. Roark Capital Group, an Atlanta-based private equity fund, for instance, is named after a Rand protagonist and as of this writing had $33 billion in assets.

Rand's influence applied not just to the boys and moneymen of today but also of the past. She enthralled Alan Greenspan, the American economist who served five terms as the thirteenth chair of the Federal Reserve, overseeing a great recession. (Greenspan worshiped Rand and even wrote essays for her *Objectivist* newsletter and was a contributor to her book *Capitalism: The Unknown Ideal*.) In a 1966 letter, Ronald Reagan wrote, "Am an admirer of Ayn Rand . . ."

Yet Rand herself also ultimately contradicted her belief system. In old age, Rand was forced to betray her mantra of self-sufficiency. After she had surgery for lung cancer in 1974, and in absolute contradiction to the values she spent her life peddling, Rand turned to Social Security—there are Freedom of Information Act documents that confirm she received the assistance. She purportedly also depended on Medicare to ease her medical expenses. Rand did not seek out the benefits herself but allowed someone to do it for her and by 1976 had grown so debilitated that a representative with power of attorney applied for public assistance. When social worker Evva Pryor was asked by Rand's lawyers to talk to her about the benefits, she said, "Whether she agreed or not [with using the benefits] is not the issue, she saw the necessity for both her and Frank [O'Connor, Rand's husband]." It was a significant step for someone who denied the value of the social safety net and said there was no such thing as public interest.

Rand's own late-life hardship and need for governmental aid was the hidden story of her personal dependence, a buried truth that like

Alger's is not widely known. Rand devotees continue to mystify her reliance on state aid, arguing preemptively that Social Security is not in contradiction with her philosophy. The chief philosophy officer (I am struggling not to put this title in quotations) of an organization named after her, the Ayn Rand Institute, rushed to defend her for relying on it. "Rand's position on the welfare state is no doubt controversial," writes Onkar Ghate. "But for critics to dismiss it as hypocrisy is a confession of ignorance or worse."

What is indisputable is that, in her old age, Rand needed both the assistance of caregivers and the state, as most of us eventually do. Her late-life condition was part of the way in which the diktats of other self-made figures tend to rest on a central falsehood and smaller lies as well.

The basis for Rand's denial of the "public interest," like Wilder's insistence on a lack of social care, lay in Rand's childhood. The self-appointed prophetess with a black pageboy cut was born Alisa Zinovyevna Rosenbaum, and from her early days her rage at social programs stemmed from having been born affluent and seen her family's comforts disintegrate during the Russian Revolution. As is the pattern for those who most aggressively assert their individualism, Rand not only invented the name Ayn Rand but also created as much as she could about her persona. This denial of her parentage and background mixed with an unfettered capitalistic impulse. Rand's biographer Anne Conover Heller depicts her as fixedly, even deludedly, sui generis: "Later, she would say that neither her family of origin nor the country she was born in had any determinative meaning for her, because they were accidental, not chosen by her own free will. She was 'a being of self-made soul.'"

Also like Wilder, Rand was dedicatedly opposed to the community-oriented bent of the New Deal, although unlike Wilder, Rand's hatred of the community-minded stemmed in part from fleeing the Soviet Union, claiming it was to visit relatives, to arrive in New York in 1926.

Her attitudes crystallized with her first novel published eighteen years later, *The Fountainhead*. Its protagonist, a nonconformist architect, Howard Roark, ostensibly displays his powerful individual will and genius by building skyscrapers and lecturing everyone. "It was a country where a man was free to seek his own happiness," Rand writes in the novel, "to gain and produce, not to give up and renounce."

The claim that all of life was either "gain and produce" or "give up and renounce" may seem childlike in its simplicity, perhaps unsurprisingly for a writer who believed most Americans were "parasites."

"I came here to say that I do not recognize anyone's right to one minute of my life," says Roark, a sentiment that is historically much easier for men and especially wealthy men to utter than for anyone else. "The world is perishing from an orgy of self-sacrificing."

One night, I watched the 1949 film based on the book, starring an exhausted-looking Gary Cooper as Roark, a master builder of modernist buildings, vicious-looking giant white structures that resembled gargantuan and sometimes jagged teeth. The setting of *The Fountainhead* was still the collectivist society of the novel, in which individualists face rejection for their excellence. The usually reliable director King Vidor shot Cooper and Patricia Neal and the other actors in sharply contrasting black-and-white, often set against an enormous New York office window (really a stage set), from which the viewer sees other looming buildings, all phallic symbols of commerce, creating a "menacing atmosphere," as one critic wrote.

The film consists of an excess of speeches about the power of one Great Man. "Great men can't be ruled," says an older male character, leaning over a terrified-looking young man in a leather chair, the second-banana character Peter Keating, played by Kent Smith, who eventually begs Roark to "save" him because "I've never had an idea of my own."

In its time, the film, deservedly, bombed. This was in part because the script was written by the absolutist Rand herself, who had been an amateur scriptwriter for years.

Again, we could sneer at the films and the books and Rand's whole position as dated schmaltz if her ideology had lost its broad influence. It hasn't, though. In the twenty-first century, Rand remains revered by many who consider her prescient, predicting everything from the strip mall to the Tea Party to what one political scientist calls our nation's current condition as a "privatized state."

"The reign of the cruel optimism of Mean Girl Ayn Rand," writes scholar Duggan, is not that she is the "presiding spirit for right-wing nationalism" but that she's actually the militant form of our worship of the market, which is far more widespread. In addition, Duggan tells us the action that we should take to protect ourselves from the noxious individualism Rand peddled: that we must "reject Ayn Rand." "After all," writes Duggan, "she rejects you."

For such a rebuke to be commonplace, we would, though, need to teach as many Americans as possible what an Ayn Rand or Horatio Alger story really means—the lies and the needs and the toxins embedded in them. Alger scholar Martel observes that Alger's "tales of dependence, intimacy, and connection are repackaged as tales of autonomy and the triumph of the self. The transformation happens before the reader's eyes, and Alger is very much in on this conspiracy." Rand had a similar trajectory. For the terms of these icons to be challenged broadly, or even undone, it would call for a mass tutorial about their hypocrisy.

After all, if every love story is a ghost story, as one novelist wrote, every Horatio Alger story is also the story of a lie.

II

**BROKERS OF
THE AMERICAN DREAM**

5

Rich Fictions

TALKING TO STEPHEN Prince is like entering a southern storytellers' convention, like having your being flooded with bourbon. His words tend toward the adjectival, like *horrible* or *righteous*. He wears colorful bespoke checked dress shirts. His skin is flushed with sun and good health, the color of shrimp caught on the little fishing boats he watches on the horizon of the Atlantic, which he can easily see from the huge ocean-facing balcony of his penthouse apartment on St. Simons Island, Georgia.

Prince is, in short, a very wealthy man, one who overflows in more ways than just financially. From the outset, the question arises, as it does with most megarich people, how did he make his money? Prince was the founder and president of a plastic gift card company, National Business Products, and made a fortune through it. In 2013, it printed one million cards a week, loyalty and gift cards for businesses like Walmart and McDonald's, all out of a factory in a business park in Nolensville, Tennessee. He is now edging toward retirement.

I sought him out because I was hoping he could explain why the wealthy could consider themselves utterly sui generis, although Prince is far more conscious of his innate privilege than many in his position. Prince believes that the Treasury Department has asked too little of him—and others like him—and that he has had it way too easy, that it's time for the wealthy to turn away from their vain attachment to the bootstrapping myth and acknowledge that they didn't go it alone.

Yet in the face of our country's income inequality, almost all of Prince's friends voted for Trump out of further self-interest, because all of Prince's friends who are right-wing "want a tax policy that protects us and our income for self-serving reasons." After all, he and they continue to make "a boatload of money in the stock market." A wooden novelty sign on the wall behind him reads "Papa: The Man The Myth The Legend," a gift from his kids. There was a bright and glittering view of not just the sea but also four thousand square feet of his window-filled apartment housing a wraparound couch, white hydrangeas in a vase, and six television screens. He had just bought an art gallery for fun, he told me, because he noticed one in his town that was set to close and he had always wanted to own one.

Prince wasn't born into this life but grew up in the loosely defined middle class: his father was a retailer, with, he said, a "sordid past" and some connections with white-collar criminals, or, as he put it, "Go back and read some Faulkner and you'll understand what I am talking about." Back in Waycross, Georgia, he told me, he was "raised in an environment that should have led me to hold politically different beliefs. What was illegal for a Black guy was not illegal for White guys. Or for my father." In addition to engaging in seamy and potentially illegal activities, his dad managed a large local franchise of the national auto parts chain Western Auto. Prince had greater ambitions than that, though. He put himself through college at North Carolina State University. Then he spent years making his money unglamorously, working for a payment processing company. In 1993, he used personal credit to found National Business Products, which started off as a corporate printing operation but eventually became focused on corporate gift cards. In 2019, Prince sold his company for "tens of millions," as he said breezily. The majority of those millions flowed to him. Prince admits that he has not always recognized his debt to the employees who helped build his companies. "This has been a real battle for me," he said.

Eventually, even though he could choose the self-made mantle, Prince rejected that label, acknowledging that from the get-go he had benefited from both the sacrifices of his hardworking employees and the confidence that's baked into being a comfortable White man in the South. He seeks to set himself apart from his rich friends, who are almost all conservative and like to emphasize their own hard work and deservedness to justify their advantage in their community. "These guys think they hit a triple, but the truth is they were just born White in the South," he said. As the economist Paul Krugman puts it, Americans use moral judgments about the wealthy and the poor as "fancy footwork to propagate what I think of as the myth of the deserving rich."

"I am thinking of how with the bush planes in Africa, how people come to clear the animals off the runway so you can fly," Prince said, offering a metaphor evidently gleaned from one of his safaris. "Someone is always clearing out the runways for us White guys."

He felt he had to do better, and he acted on that feeling, supporting youth programs and local Democratic candidates and, perhaps most importantly, publicly speaking out about taxing the wealthy at higher rates. He joined the Patriotic Millionaires, a group of high-net-worth individuals whom you will read more about later; they are leading a rebellion against how the richest Americans are taxed and raising the collective consciousness about it. "Tax cuts for the rich at the expense of the working class is not a sustainable system for our economy," he writes in an op-ed for the *Guardian*, one of his many media appearances as he participated in a pro–tax-the-rich campaign that his wealthy peers would say was against his own self-interest. "My peers do not, by and large, take their tax savings and reinvest them in higher wages and job creation—all the things that might help stave off a recession. Instead, they tend to keep it all to themselves because our tax laws allow them to."

What Prince has been offering is worthy, although perhaps a weaker structural reimagining than we might wish to see. After all, just hoping the wealthiest will start to think differently is not enough—we don't just need better multimillionaires to correct our social problems with their occasional impact investing or GivingTuesdays.

Yet sadly, envisioning a broader view of society while still existing within this milieu is an accomplishment. Many of the richest today are not even attempting to be as helpful or as self-aware as Prince, or even sustaining a noblesse oblige pose. They may instead tend to welcome the fact that the game is rigged in their favor. Their ambitions may also be increasingly interplanetary, as expressed in this remark by Paul Orfalea, founder of Kinko's copy shops (now FedEx Office): "Well, one day I'd like to go to the moon and look at the planet Earth and say, Wow, there's part of my portfolio."

═══

This rich-person formula might seem innocuous at first: you work hard, you deserve your fortune. That might be true if their fortunes weren't so often inherited and if disparity weren't so great.

Despite evidence to the contrary, most Americans still think their society is fairer toward them than in fact it is, as if playing unwitting straight men in a cringe comedy, butts of unfunny satire. According to one survey, Americans believe the rate of reaching the top from the bottom is *12 percent,* not 8 percent. (There's only an 8 percent probability of a child born to parents in the poorest fifth of the nation reaching the middle class, or the wealthiest fifth of the population.)

Americans also believe that CEOs earn on average just $1 million a year. In truth, CEOs at the top 350 US firms made on average

$14.5 million in 2019, thousands of times more than most of those who work for them. And while the actual pay ratio of CEOs to unskilled workers in America earlier in the last decade was 354 to 1, that calculation far exceeds what ordinary Americans estimate, which is a 30-to-1 pay ratio. That in turn far exceeds a more desirable ratio, which is 7 to 1, as Sorapop Kiatpongsan and Michael I. Norton found in their 2014 study "How Much (More) Should CEOs Make? A Universal Desire for More Equal Pay."

The highest CEO pay during the pandemic years went to Paycom CEO Chad Richison, who made $211 million. One might also cite in the Pandemic Book of World Records Hilton CEO Christopher Nassetta, whose yearly compensation was $55.9 million, 1,953 times as much as the company's median worker pay of $28,608. As Barbara Ehrenreich writes, in what I believe is half-seriousness, Americans can't actually "afford the rich, who drive up costs for everyone and require huge amounts of cash to sustain their lifestyles."

In short, income inequality isn't something that you can see clearly when Mark Zuckerberg goes on holiday, paddleboarding at his $59 million compound: it involves an internalized illusion held by those who are victimized by it yet still believe what I call its "rich fictions," and these delusional fables are part of what keeps ordinary people locked in the bootstraps belief system.

In political life, rich fictions are evident when the wealthiest claim that their own and other rich people's "self-made" status is what has gotten them to the top of the hierarchy. This is not usually the case. The group United for a Fair Economy found in 2012 that over 60 percent of the *Forbes* 400 list of the richest Americans were already well-off when they began their careers. They might have had an upper-class upbringing and then the gift of start-up capital from a wealthy family member. Around the same time, French economist Thomas Piketty estimated that roughly 60 percent of America's private wealth was inherited rather than the result of the much-ballyhooed hard work. At the same time, Americans think their odds of success, and of rising from the bottom to the top, are much higher than they are, and they have greater expectations based on more slender reeds than those relied on by their European counterparts.

When running in 2012 as the Republican candidate for president, Mitt Romney announced at a town hall in Ohio, "To say that Steve Jobs didn't build Apple, that Henry Ford didn't build Ford Motors, that Papa John [Schnatter] didn't build Papa John's pizza . . . [is] not just foolishness. It's insulting to every entrepreneur, every innovator in America." (Romney might have considered retiring entrepreneurial heroes, as Robert Reich suggested we should do in the 1980s, exchanging the archetype of the Executive Alone with a more honest model that prized teams of coworkers.)

In Regency-style rooms or modernist breakfast nooks, in private dog runs, in whole nightclubs rented for their college kids' birthdays, these rich families also now wield a wildly historic level of privilege over the rest of us: in 1973, America's top 1 percent of income earners made 9 percent of the nation's income, while today they make almost 25 percent.

During the pandemic, this tendency to let the richest Americans run roughshod over the rest of us and continue to create alienating inequality reached a fever pitch. While half a million people had died during the COVID-19 pandemic by February 2021, and forty million were out of work within the first two months, billionaires, along with many mere millionaires, continued to make swollen profits from their town house basements or private planes, executing trades or ordering their managers about. Their earnings while so many were suffering exposed the unequal sacrifice Americans were making during the crisis. Billionaires stashed away $434 billion during the hard times between mid-March and mid-May 2020. In 2021, the collective gain of the richest was a staggering $1.6 trillion, a swelling of 55 percent, while millions of Americans were at risk of losing their homes. After the debut of the annual *Forbes* billionaires list in April 2021, the *Washington Post* wrote, "Amazon founder Jeff Bezos, with an estimated fortune of $177 billion, topped the list for the fourth year running," while Elon Musk, the Tesla chief executive, "came in at No. 2" at $151 billion and Zuckerberg, CEO and cofounder of Facebook, came in fifth with $97 billion. There was little guilt in evidence among those who were making out like bandits during the pandemic. All that seemingly mattered was that their own metaphorical boots continued to be the newest and most expensive and that they stayed healthy and safe: some of the ultrarich scampered off to their New Zealand farms, to Scotland, or to competitive frenzies in formerly neglected towns to nab the rare estate, getting their toro salmon and Pelotons sent to them, their personal needs met by "quarantine assistants." Some of them did the equivalent of dropping out, where their "dropping out" was a new form of elitism; that they could check out from their ordinary lives only showed their riches and status.

I kept waiting for widening inequality to be understood as what psychiatrists now call a "moral injury." I was time and again surprised that many were so subdued in their response to this outrageous inequity, partly because, as public health specialist Gregg Gonsalves tweeted in 2020, "individualism and bootstrapping has made us absolutely complacent," even during pestilence, a society "that fails to rise up." Even our indignation has been tampered with without our knowing it.

The powerful tend to disseminate ideology—to those who work for them, vote for them, and invest in them, through everything from political ads and business cable shows to think tanks—while the rest of us are coerced to absorb their storytelling.

The rich fictions about being self-made or being more worthy involve the megarich both covering up their dependencies and blaming normal people for needing assistance. If they are libertarians, they are also often keen on transferring what in finance is called risk onto ordinary individuals rather than letting at least some of that risk be held by state or federal governments or employers. Their stance can often be hypocritical around self-reliance as well. For instance, Musk regularly rails against government subsidies, while his corporation, Tesla, benefits from them in the billions, as well as from tax credits. According to Subsidy Tracker, Tesla received nearly $2.5 billion, a mixture of eighty-two federal grants and tax credits as well as twenty-eight state and local awards.

Yet Musk continues to reject social programs for other people. "Another government stimulus package is not in the best interests of the people imo," he tweeted, irksomely, during the first summer of the pandemic, while his companies continued to benefit from the generosity of the government. (Musk also argued against protecting citizens from contagion, saying early during the coronavirus lockdown, "You should be allowed to do what you want," and insisting that he keep his California plant open during the pandemic, imperiling his workers.)

That the rich deserve their money is a fiction that crosses party lines, though—it knows green more than blue or red and adheres to the sort of rich person who claims to believe in good government as well.

For instance, when Michael Bloomberg got on the 2020 Democratic debate stage in Nevada, preening that he had made his billions on his own, handily forgetting the many thousands of workers who got him where he is today or the entire American infrastructure of public education, government-funded research, and even the building of the internet that was ultimately responsible for his success. He credited his estimated $64 billion net worth to his own labors, saying, "I worked hard for it." (I think of what Bloomberg did on that stage as "rich mouthing," the opposite of the idiom "poor mouthing," where one flexes one's own wealth, erasing the contributions of those who got you where you are—followed by the expectation of applause. It's a surprisingly common move.)

Luckily, Senator Bernie Sanders of Vermont was alert to this strategy and as a result was well prepared to take Bloomberg down at that debate, retorting: "Mr. Bloomberg, it wasn't you who made all that money. Maybe your workers played some role in that as well." Sanders knew to publicly puncture the gospel of individual success by those at the top of the income gradient. He was, however, working against a catechism that has kept the wealthiest 1 percent (and those directly beneath them, aiming upward) making bank, while the rest of Americans were being punished by a social order stacked in their disfavor.

———

The so-called self-made are not just male. They include some of the most privileged women, too, showing that even feminism could be appropriated under the bootstrapping mantle.

Yet a majority of the wealthiest are not strivers. The 2019 edition of the *Forbes* 400 list showed that most of these affluent women were heiresses, emerging from American dynasties like the Waltons of Walmart. In fact, only eleven self-made women made the list at all.

Nevertheless, rich fictions tend to animate and suffuse how these women talk publicly about the role of individual effort in success. As billionaire "power feminist" Sheryl Sandberg puts it in her 2013 bestseller *Lean In*, turning what should be self-help into a text that foments ordinary readers' self-doubt: "I continue to be alarmed not just at how we as women fail to put ourselves forward, but also at how we fail to notice and correct for this gap." There was little systemic critique in Sandberg's books or most of the girlboss books.

Even lower-brow versions of this hectoring tough love also flowed out of other best-selling entrepreneurial self-help books by women like Rachel Hollis, who wrote *Girl, Wash Your Face*. Hollis writes, sounding accusatory rather than friendly, "I absolutely refuse to watch you wallow." The girlboss, as Hollis and others were colloquially termed over the last decade—a term that, as I write this, is already an artifact, a cute neologism coined by the founder of the online women's retailer Nasty Gal, Sophia Amoruso—was a feminized rich fiction. Girlbosses included photo-ready female impresarios like Audrey

Gelman, a sylphlike former publicist who raised $100 million to co-found The Wing, a chain of social work spaces in and around New York City. An expensive membership-only franchise that catered to women, The Wing described itself as "Empowering Women Through Community." Her image was augmented when her wedding appeared in *Vogue* with bridesmaids in vintage-clothing-inspired outfits designed by the creative director of J.Crew. There was Yael Aflalo, of the clothing line Reformation, credited with creating "Every It Girl's Favorite Brand" by a fashion site that oozed that her designs could be found on "the coolest girls on your Instagram feed." Girlbosses might be photographed on their laptops, with their desk chairs the pale red or deep pink color of a uterus. In its most mass-market manifestation, a girlboss culture consisted of cutesy-poo volumes like *The Girlboss Workbook*.

Despite the newfangled, attractive packaging and the soupçon of girl power, girlbosses paired nicely with preexisting systems of power. They, like other purveyors of rich fictions, tended to insist they had built a business from scratch by themselves—or that their model of solo financial success was somehow fresh and new and millennial. They are faithful to success itself without acknowledging that success in a country with such limited class mobility isn't generally the result of speaking up in the conference room, or not wallowing, as much as it is the product of wealth, skin color, access, and a slew of related advantages. They tell their female consumers and readers that success requires working within corporate or political structures rather than making attempts to change them. In the process, they filled millions of women's heads with the idea that they could easily get a greater share of that power and security if they simply tried harder—and if they didn't succeed, well, it was their own fault.

These influencer girlboss women seemed to be a new outcropping of the "feminist fallacy" where "images of powerful women in the media" imply empowerment of all women, as Andi Zeisler writes. The notion of the "trickle-down" corporate feminist can be a rich fiction all its own, even though it was an archetype whose logic seemed more fictive than usual during the pandemic. As the scholar Tressie McMillan Cottom puts it, "Trickle-down economics wasn't the best experience for people like me. You will have to forgive me, then, if I have similar doubts about trickle-down feminism." In other words, some women in America—maybe, sometimes—got to participate in the bootstrapping story: most working women did not. Cottom and other feminists, among them Black feminists who fought upstream for their place in the firmament and had to sacrifice to do so, didn't relate to the girlboss story line or even the idea of female workers "having it all," which Cottom observes "is not a feminist theory of change."

The girlboss was sometimes fictional in other ways, as when twenty-one-year-old reality TV star and model Kylie Jenner agreed with press reports that she was youngest "self-made" billionaire due to her cosmetics company rather than her Kardashian family–fueled fame and lucre, from the reality show *Keeping Up with the Kardashians,* starring Kim and siblings. Hadn't she spent her youth in a California manse that millions of people saw on television long before she started a corporation?

"There's really no other word to use other than 'self-made,'" she said, defining herself, ignoring how illusory the category was, "because that is the truth."

By 2020, however, the tide had turned. The girlbosses had an unapologetic relationship to wealth and luxury. In the wake of a wave of activism led by Black Lives Matter, their actions were suddenly exposed to greater scrutiny (and, admittedly, often a far closer examination than men occupying the same station would receive). As a result, a number of the women associated with girlboss-ery fell—or were pushed—off their perches. Their number included Gelman, Aflalo, and *Girl, Wash Your Face* author Hollis, the archetype that specialized in obliviousness toward others' realities, to the point at times of being exploitative toward employees, fans, and even consumers. Their descent was a recognition of the deceit and bias lurking within the self-made myth, although I couldn't help but wonder whether, if these women were "boybosses," they might have been spared.

Gelman's plush velvet-and-brass bootstrapping form of feminism was derided by some of her employees as not just self-interested but perhaps racist as well. Aflalo's brand also came under attack. The designer for Aflalo resigned in 2020 after she was pilloried for traumatizing her staff, including the brand's flagship store manager: multiple former employees accused her of mistreatment. And then Hollis, who had built her online presence into a mini-empire for self-improvement, was accused of lying about her personal life, appropriating Black culture, and making classist remarks.

What was overlooked in the apologies of the girlbosses who were exposed and dethroned was that even with them erased, the belief system that had raised them up was more or less intact. That some of these women had been professionally canceled didn't move the founding myths that had created them and would continue to enshrine others. Like many of the "self-made" men at the highest echelons, these younger capitalist queens had been so busy getting theirs that they managed to lie to themselves about their own dependency on everyone else. And while some particular rich fictions were exposed, the corporate feminist archetype of which the girlboss is a more recent, brightly hued iteration remained more or less unscathed, along with the blame-y imperative to climb the proverbial ladder as fast as you could or be kicked off on its lower rungs.

=====

Rich fictions are a very old story. The tenacity of that narrative is part of why academics have long tried to make sense of why people give more credit to the powerful than the rest of us and judge the less powerful more harshly than they should. In fact, one of the major research studies of this is more than fifty years old. It's called the "just-world hypothesis."

It's the idea that seeks to explain this fallacy: we desperately try to see the world as fair, even in an inherently economically unfair country such as ours. That leads us to go to extreme lengths to protect our sense of order and justice.

The just-world hypothesis was developed by social psychologist Melvin Lerner in the mid-1960s. It is described by the American Psychological Association as the belief that "what happens to people generally is what they deserve. In other words, bad things happen to bad people, and good things happen to good people." (Lerner himself didn't think the "just world" theory was true—he subtitled *The Belief in a Just World*, his 1980 book, *A Fundamental Delusion* for a reason.)

He came up with it after he noticed something strange about the psychologists at the hospital in Kansas he worked out of. He felt that the shrinks were physically rough with their mentally ill patients and used cruel and derisive language to describe them. Though they were meant to be members of a caring profession, Lerner felt that these professionals actually despised the people they were supposed to help. He wanted to know why.

So Lerner began a study, the first leg of which was completed in 1966 at the University of Kansas. Lerner and his assistants had seventy-two college women at the university gather "in a waiting room a few minutes before the scheduled experiment." Groups of four to ten women were joined, in the words of the paper that described the original experiment, by the so-called victim—a girl of their own age dressed like a student. The subjects and the victim then accompanied the experimenter to the "observation side" of a mirrored testing room. The women proceeded to watch a videotape of that other woman, also called a "confederate" in the paper that resulted from this research, being "strapped to the 'shock apparatus,'" attempting to learn nonsense syllables. During this process, the victim received several apparently painful electric "shocks" from a researcher named Dr. Stewart when she answered incorrectly. The victim twisted in agony and exclaimed, which Lerner noted was "a very effective performance."

After viewing the tape, the students were given a range of instructions depending on which version of the experiment they had been assigned. The result: most of the seventy-two women who had watched a stranger receive shocks characterized the woman negatively. Lerner and his colleagues concluded from the actions of the study's participants—and those of people generally—that there is a strong need to believe that other people get what they deserve. Inspired by the experiments of the controversial social psychologist Stanley Milgram, they were driven to understand "how societies which produce cruelty and suffering maintain even minimal popular support." Lerner observed that we cling to our belief in a just world, as we adhere to the bootstrapping myth, to help our systems of value appear "stable and orderly."

"If observers can attribute the victim's suffering to something the victim did or failed to do, they will have less need to devalue his personal characteristics (other things being equal)," Lerner and scholar Carolyn H. Simmons write. "The observers' belief in a just and predictable world will not be threatened."

Their findings also apply to the ways in which we internalize and propagate rich fictions. Humans need to see the world as fair, just, and predictable. This desire can be so psychologically important that it is easier for us to watch a woman be punished with just the thinnest of pretexts—and view her as warranting her abuses—than to believe her treatment is random. To keep ourselves sane, we may twist the truth to make our own existences more bearable—we are poor or alone because we are undeserving; that stranger is ill or failed because they are ineligible; that millionaire heiress has a Tesla or makes bank as an influencer because she is a person of merit. This confusion keeps the majority from recognizing that this is not always the case.

The just-world theory helps me understand why so many assume that the poor are not hardworking and even explains resistance to student-loan forgiveness and continued stigmatization of the forty-six million borrowers who collectively owe nearly $1.75 trillion of it. The reasoning is that if they are meritorious, they should be wealthy enough not to need forbearance—and if they aren't, writing off their debts would taint those who had paid them back as suckers.

Similarly, the University of Kansas's "just world" study participants unthinkingly believed that the apparently shocked victim was somehow deeply—perhaps morally—flawed. We also project those same moral blemishes on others to rationalize income inequality and moral deservingness to explain wealth. (This is part of why some people feel they have a chance to win, believing that even if a system creates inequity it is nonetheless "fair.")

In later decades, other related theories echoed that which was laid out by the just-world hypothesis. There is "status construction theory," developed largely by Stanford University sociologist Cecilia Ridgeway, who argues that social hierarchies can help us understand how one group tends to, say, be broadly seen as more important or more competent than another. Ultimately, people who have been dubbed lower status wind up giving in to the demands of the higher status people. During the pandemic, status construction theory helped me better comprehend how many privileged people believed they *should* be protected from contagion.

More recently, Paul Piff, a social psychologist at the University of California at Berkeley, asked similar questions about ideas of merit and the American Dream. For one study, he used a rigged game of Monopoly where players were defined as either rich or poor by flipping a coin, with the former getting twice the amount of cash as the latter at the start. Predictably, the poorer player would almost always lose. Yet what was striking was that the richer player very quickly believed they were owed their advantage: their win, in their minds, was due to superior skill, not initial luck.

=====

Over fifty years of research was still unlikely to move those who subscribed to rich fictions to view their achievements as relative. Stephen Prince's friends were, after all, still convinced that personal virtue was what made them who they were. (I did wonder why he was still friends with them.) His friends didn't tend to allow that "a lot of success is luck," as Prince said. "I've been lucky. People don't admit to being lucky, as it takes their genius away, but I am here to tell you that rich people are not so damned smart."

How Prince wound up with such a different take and such a divergent attitude from those who shared his background, he still can't quite explain. Was there something we might learn from his example? Could we try to convince other rich people to follow his well-intentioned (and well-tended) path?

Prince took one fork in the path earlier in his life. He remembered when his own father applauded a cop after the police officer jailed a young Black man for refusing to sit at the back of the bus, and he took his father's actions to heart, most negatively. This was a flashpoint, he said, for rejecting both that kind of racism and the illusion of merited White wealth. At college he further departed "from traditional Southern morals and values" and soon felt like "a weird bird" to be living there at all.

Another defining period, Prince said, was the 1980s, when he "saw the inequality that was developing and knew that trickle-down economics were a lie. I know how rich guys are. They say, 'I am going to make a whole bunch of money.' They don't say, 'I am going to pay lots of people well because it's the right thing to do.'"

"My generation has screwed us up," Prince told me. "Americans are a good people, but Reagan kicked something up in 1982, gave us all a pass to be selfish assholes; I don't think that's who we really are." Prince isn't wrong: one of Reagan's campaign slogans was "government is not the solution to the problem; it is the problem," and as Nancy Reagan put it, "Somebody will say, 'We have a problem,' and the immediate response is, 'Well, can't government do something about that,' instead of, 'Can't I do something about that?'" Also in 1981, when Reagan took office, any married couple's estate that exceeded $350,000 was taxed—now that threshold is over $22 million.

Prince said, "I am for any kind of tax, including the wealth tax."

This attitude among the rich toward paying their taxes, combined with just how useful those taxes would be, led Prince to spend his free time during the last administration telling media outlets that Trump should raise taxes for rich Americans like him.

As of this writing, this was not to be. The rich Americans were figures of fun for their obsession with going to the moon, making inequality galactic. There were small rays of hope for the rest of us when President Biden and even Democratic senator Joe Manchin of West Virginia and obdurate moderates signaled that they felt better taxing the rich than increasing the deficit.

For Prince, what spurred on our mass-misguided American faith in self-sufficiency was race and gender.

"I've been ragingly successful in life, and part of why our runway was so clear to take off into the stratosphere of success is simply that I am a White male," Prince mused. "As that is so, I should be handicapped like a jockey at a horse race: if the jockey is too light, they put more weight in the saddle; the same thing should be true in our society."

Reaching for another metaphor befitting his good-old-boy origins, Prince added, "We White guys have all the face cards and all the aces—there is no way for you to win."

6

The Self-Made Voter

There is no such thing as society.

—MARGARET THATCHER

WHEN RANDY ROECKER and his partner, Linda, decided to go to a Trump rally in October 2020, they weren't expecting to get emotional. Nevertheless, the feelings flowed, said Roecker, especially when "Ave Maria" blasted and they properly took in the giant American flag that flapped in the wind while Air Force One flew overhead to top it off. Linda even cried. "We were so inspired—it was an extreme feeling of patriotism," Roecker said. "The music, oh my God, and the people, they were so friendly. It's conservative out here in America: most of my neighbors supported him." Roecker was stirred by "the thousands and thousands of people" at the rally. And he was "thrilled" in particular to be seated so close to Trump himself.

Roecker was then a fifty-seven-year-old dairy farmer living in rural Wisconsin outside the village of Loganville, population 300, where the closest place to shop is the 9,500-person town of Reedsburg. Roecker first found common cause with Donald Trump in 2016 when Trump was running for president; foremost among his reasons: that Trump defined himself as a self-made man, even though the candidate, contradictorily, was born rich.

Another rationale for his interest was that Roecker had grown irritated by the mandate to wear masks and the lockdown. "We know the virus is out there, but we're not just going to shut our lives down," he said. "We feel like they are wanting to take our rights away."

Trump's dismissive attitude toward the virus supported the local view that precautions were not always needed. Even in relatively early COVID days, Roecker's neighbors—mostly farmers and local businesspeople—felt that the safety measures were trampling on their individual liberties. In addition, "People are scared to death of the vaccine," Roecker said around the time that the vaccine became available to those his age. (He wasn't zealously antivax, however: Roecker's parents would be early recipients, and Roecker himself would get the shot after his cousin, a health-care worker, encouraged him to set an example.)

With Trump, Roecker experienced twin sensations: of being understood and identifying with someone with power. There was the promise of freedom from the social restrictions he and his cadre had felt restrained by, and Roecker also felt that Trump was akin to the people who had reared him, who were not as educated or fancy in their speech as politicians tended to be. He chalked that up to the then-president's "fresh businessman's perspective," along with the rough clarity of his words. That Trump was sometimes clumsy or even brutal in his utterances was actually "a selling point. He's a builder who talks like a builder. That's what we understand in the heartland, not some coastal American talking!" As if to illustrate the difficulty of translating what is trendy political language in some quarters to the populace at large, Roecker asked me to define the word *misogyny*, which he found unfamiliar, and said he found the use of such words by Democrats and their allies alienating. Meanwhile, Trump understood people like him, he said, and he believed he understood Trump.

Roecker needed to feel mirrored and sympathized with, because he had suffered a great deal at times. He had been depressed to the point of hospitalization a few times, years back; much of his melancholy he attributed to the financial distress of being a farmer. After hearing of a spate of farmer suicides that had been caused by a similar agitation, Roecker created a foundation to spread the word about that epidemic. In an interview with a reporter around that time, he said of his own experience: "You have this burden that you carry. I kept feeling all the time that I was a failure, that I had let everybody down." He had even gone to his barn and considered stringing himself up from the ropes in the hayloft, thinking that this was one of the least painful and most metaphorically apt places to end his life.

Throughout our conversations, Roecker was gregarious to the point of boisterous, belying his own experience of deep sadness. As he sat on a bench on Main Street in Reedsburg, he told me how the last decade had been particularly tough for him and his friends in the farming business. "It's everyone," he explained, his voice rising, "the milk but also the hog and beef, corn, and soybean farmers." Roecker estimated that on average farmers in his town were each losing $30,000 a month—despite quarantined consumers who were now "staying home to cook with butter and to drink more milk." He doesn't mention the $46 billion in direct government payments to farmers in 2020, which has been called an agricultural bailout, breaking previous farm-subsidy records, but he said he would eventually apply for a federal loan designed to help small businesses like his that had struggled during the pandemic.

With his parents and adult children living on the farm, Roecker felt trapped into continuing to run it, even as it sustained more losses. "Who is going to buy a business that is losing money? Its only value is the land," he said, of his farm and his hundreds of cows. He said he would love to leave this vocation—to "run away from home" and forget about all of the "paying down and borrowing back" that he was undergoing to avoid bankruptcy.

Roecker, like many bootstrapping Americans, especially those who supported Trump, feared losing his status and foothold. He worried about being forced to let go of what he still earns and his farm being taken from him. Like his parents and kids, Roecker had been a dairy farmer all his life. "My grandfather started our farm with ten cows, pigs, and chickens. My dad was milking fifty cows," he said. Roecker has six times that, yet he lives in fear of losing them to "venture capitalists that want to invest in robotic milkers." He voted for Obama in 2008. Then after "numerous bad years" that he largely blamed on "milk prices set by the government," he went for Trump in 2016 and again in 2020. At that time, milk prices were going down again, and, as if by some cruel biblical decree, the cows' feed costs were simultaneously going up. "My mom is crying today as she had no money to pay the electricity bill for the dairy," he told me.

———

I started talking to Roecker and other current and former Trump voters in November 2020, after the election had gone to Joe Biden but before Trump conceded, in truth only sort of. To better understand how the self-made narrative had upended politics, I also waded through stacks of books about "the success ideal," books with titles like the 1965 *Apostles of the Self-Made Man*. I followed that up by scrolling through recent surveys on the subject. That was when things got more interesting.

The most striking one was conducted by Jared McDonald, a postdoctoral fellow at Stanford University. The survey asked Democratic and Republican voters how much a candidate's defining themselves as self-made influenced their voting. McDonald and his collaborators, David Karol and Lilliana Mason, were also trying, more broadly, to assess "the effect of biographical knowledge on voters' assessment," as they write. It turned out that voters were very influenced by what they believed to be candidates' biographies, particularly when they perceived them as having successfully bootstrapped their way out of humble origins. In 2018, when the researchers asked, "To what extent were you aware that Donald Trump grew up the son of wealthy real estate businessman Fred Trump, started his business with loans from his father, and received loans worth millions of dollars from his father in order to keep his businesses afloat?" they learned that quite a lot of the voters did not know this.

Many also didn't know that Trump was raised in great wealth. According to the survey's authors, "This misperception increases support for Trump, mediated through beliefs that he is both empathetic and good at business."

This gap in understanding was partly due to decades of Trump's doctoring his story, using his celebrity to enforce this notion of himself. In his youth, he had been heavily influenced by Dr. Norman Vincent Peale, the Trump family's pastor and author of the best-selling book *The Power of Positive Thinking,* who believed he was "God's salesman," as his biographer Carol V. R. George called him. His theology was to encourage self-confidence verging on personal arrogance. Peale's first rule was to "formulate and staple indelibly on your mind a mental picture of yourself as succeeding."

Self-made rhetoric could also hide the truth: that Trump had been an heir accruing debt who liked nightclubs. Also, his family had benefited from governmental generosity—from not paying taxes. And Fred Trump had received financing from the Federal Housing Administration during the Depression so he could build more than two thousand homes in Brooklyn between 1935 and 1942 alone. Then, later banks bailed Trump the Younger out. As Brian Miller and Mike Lapham put it in their book *The Self-Made Myth: And the Truth About How Government Helps Individuals and Businesses Succeed,* "The survival of the [Trump] company was made possible only by a bailout pact agreed upon . . . by some 70 banks, allowing Trump to defer on nearly $1 billion in debt, as well as to take out second and third mortgages on almost all of his properties." The pretense became even clearer in the final year of Donald Trump's presidential term, when it was revealed that he had paid almost no federal income taxes for years on his hundreds of millions of dollars.

At the same time, Trump implied that he believed inherited wealth to be a sign of inborn superiority, crediting his success to his "good genes" or "the right genes," supposedly better bred, on a cellular level, than others. As one of Trump's biographers puts it, the family "subscribes to a racehorse theory of human development. They believe that there are superior people and that if you put together the genes of a superior woman and a superior man, you get a superior offspring." Trump simultaneously indicated the opposite: that he had gotten where he was on his own and "earned" his good fortune in this way as well.

Yet these details of Trump's life—that he was profoundly privileged and subscribed to a genetic worldview in which those born rich were biologically elevated—were lost on a bunch of American voters who were being surveyed.

At some point while surveying, the researchers offered accurate information to the respondents regarding how Trump's father influenced his son's career. The information the scholars shared caused "respondents to rate the president more negatively on both empathy and business ability. These findings suggest that correcting information about candidate characteristics can change the minds of even loyal partisans," according to the study. Voters learned their candidate was not self-made. The answer, said McDonald, was "not only were people not aware of his growing up wealthy, when we gave them that information it changed their views of him. Republicans' faith in Trump went down by 10 percentage points."

If the fact that a candidate is *not* self-made has a political effect, why did such a revelation about Trump—shared widely by respected news outlets across the country—do little to undermine his popularity with his base in the 2020 election?

"Some Republicans knew deep down that Trump's not a self-made man," the researcher explained, but at the same time they "wanted to believe he'd had similar experiences to them that he's overcome" and then that "he's shown us the way." They convinced themselves that "he understands the problems of people like me," said McDonald. By persisting in their beliefs that Trump was a "business savant who took a small loan from his father and turned it into a massive empire on his own; that his family was poor, or working-class," said McDonald, these voters were able to make their identification with Trump feel authentic rather than strained.

According to this reasoning and research, the self-proclaimed self-made sector—neither poor nor rich—might well continue to stand by similarly nativist and populist conservative candidates as they inevitably emerge.

Their passion for politicians who purport to be self-created did not arise in a vacuum. This variation on the myth, with its hatred of "losers" and "victims"—those who express need or vulnerability—was a vulgar, hardened version of impulses that were already present across decades of our political life. This particular zeal had gotten more endemic, though, in recent times, as a wide swath of Americans felt humiliated by elites and forgotten.

Katherine Cramer, a political scientist at the University of Wisconsin at Madison, whom I have relied on for her perceptive readings of the nonurban state of mind, observed to me, "The self-made thing is the crux of what people think of Trump in rural America." She had studied Trump supporters for the 2016 and 2020 elections in Wisconsin and said of them, "They see him as standing up against the type of people who look down on them." Cramer said she believed that these voters think to themselves, "We can figure out what's best, not what the government is telling us; we should decide for ourselves."

In her book's pages, these Trump supporters decry Democrats for "laziness." (Her agrarian Wisconsin subjects also thought of blue state types as invaders looking to pick up real estate—appropriative urban prospectors around every weather-beaten barn door.)

In states like Wisconsin, Ohio, and Michigan, a hearty number of union members also went for Trump in 2020. Had they also bought into the fantasy of the self-made man? Yes, according to Bob Kemper, the grievance chairman for United Steelworkers (USW) Local 1299. Kemper represented Great Lakes workers in Michigan, and he identified with an earlier Democratic incarnation of the USW, even though some of the membership had shifted away from this affiliation. Kemper now represents plenty of union workers who are Trump fans and true bootstrapping believers. (While support from union households in Michigan and Wisconsin was ultimately crucial to President Biden's victory, the majority of union households in Ohio and Pennsylvania voted for Trump in 2020.)

Most of the men Kemper and his ilk worked with earned at least $80,000 a year, despite often having partial college educations at best. When laid off, they had relied on benefits like SNAP for food when necessary, Kemper noted, yet they tended to believe that federal support was a pure form of socialism, something they opposed.

Many were Trump supporters who had to believe in the narrative that the Democrats are serving the poor. "They are just parroting what they hear," he said of those with this stance. "They'll say you shouldn't live off the government, but as soon as they lose their jobs, they want as much fucking money as they can fucking get."

Kemper himself now made "an excellent living—I left school in eleventh grade and got a GED and got lucky. I had the opportunities in manufacturing. I mean, I may come across great in a fucking interview, but my education isn't that fucking good." Before joining the union, he had been a barback, labored in a steel warehouse, and done maintenance work at a chemical company. Then he joined the USW and worked on the floor as an electrician before being elected as a divisional rep. In April 2018, he was elected to this chairman post.

That was when it hit him that "we make a comfortable living for the level of education we have—and guys don't want to realize how unmarketable you really fucking are," he said.

Democrats, to secure their advantage over Republicans with this crowd, would need to expose that one of their sins—Trump's sin—is that the GOP is not in the union man's corner, said Kemper, and that they were "screwing them over." Of course middle-aged steelworkers *were* being told this truth again and again and again, but often it was via papers they didn't read or on radio and TV shows they didn't tune in to. There were also those who just refused to believe it.

This gave him special insight into the men he served, Kemper said. He added that he could easily imagine why they backed Trump, leaving the former union stronghold of the Democratic Party behind. At the root of it, he thought, was their taste for Republican bootstrapping rhetoric. It was also out of a fear of losing what they have. Even though on paper many were relatively affluent, their limited educations and the fact that they were older meant that they would be almost unhirable in their communities if they were laid off. Before many of the steel plants started to shut down, said Kemper, these men "sat high, the head of your table, the family member who makes all this fucking money."

Scholars have called this a state of "loss aversion." Studies by cognitive psychologists, including landmark scholarship in 1992 by Amos Tversky and Daniel Kahneman, have found that losing—money, status, or anything of value—is twice as powerful as the joy resulting from gaining something beneficial. This creates a bias against loss. And that fear of disappearance explains why some choose to avoid losing ground over making gains and why Trump supporters who earned an average of $72,000 a year in 2016—a middle-class income—still had a fear of evaporating social status. As the policy analyst Heather McGhee writes in the *New York Times* in early 2021, Trumpism is associated with the "stinginess of traditional conservatism, along with the fear of losing social status held by many white people." These guys are "crying about their jobs when they talk to me," Kemper said. "Now you have lost this status within your social structure. Look, people my age, the last thing we want to do is go out there and find another job."

This antagonism to even the appearance of diminishment is not the province of Trump supporters alone. In reporting my last book, I met progressive, loss-averse middle-class people, so I'd been thinking about this condition in its various forms for nearly a decade. Some were riven with anxiety about their futures and saw themselves as existing in a self-interested society.

Kemper also saw why some of his men had felt disregarded by the Democratic Party and gone over to Trump. (I thought of Andrew M. Cuomo, the now fallen former Democratic governor of New York, when Kemper said this, and how during the pandemic he'd become an unlikely salesman, in his television addresses, for the power of the collective over the individual, and how that, too, was revealed to be lip service.) Democratic politicians called on the union every four years, Kemper said, and "paid the unions lip service, and once it's done they are pointy-head fucking nerds with no relation to working men down here."

Today, two-thirds of American adults lack four-year degrees. At the same time, a person without a college degree encounters more obstacles and indignities than graduates, including earning less now than in 1979, adjusted for inflation.

===

These Trump supporters and former supporters, and the scholars studying his popularity, led me to wonder what could have been if newscasters across the board had, before the 2016 election, reminded viewers of, say, the falsity of some of his claims. Might we have had different results? This exposure *had* come on CNN and in the pages of the *New York Times*, which debunked Trump's achievements and exposed him for ripping off his vendors, but certainly not in many quarters of media.

Would that have changed things for Kristin Cole, then forty-five, a former nurse turned stay-at-home mom near Scranton, Pennsylvania, who supported Trump in 2016, as her husband did then and again in 2020? Cole's husband worked at a power plant, where he earned a healthy $120,000 a year, and he and his buddies and coworkers were passionate Trump endorsers. She seemingly twinned Trump in her mind with her husband and his friends, the "hardworking" guys she loved. She also was an unapologetic booster of bootstrapping who spoke with hostility of the "kids with graduate school debt." "No one is telling you to go to these fancy schools! I went to a cheap nursing school. You have to take personal responsibility." (Her parents, who were more affluent, wouldn't pay for her college: "They cut us off. My father thinks that money is his.")

Cole said she believed that some millennials expect more from their lives and society than even she and her husband got. These are, she said, the "younger kids who are complaining that 'I'll never buy a home or live an American Dream.' No one helped us. We bought homes on our own. All of this on our own with hard work.

"I believe in being self-made: that was why I married the guy I did," Cole continued. "My daughter sees her father go to work every day, never missing a day of work, never taking a sick day." As an example of how do-it-yourself the couple is, she said that when they remodeled their home, "We put huge windows in ourselves."

Her husband had grown up poor, as she put it, "and he left his family on his own to go make it for himself—he is prideful." Her spouse is such an ardent employee, she told me with pride in her voice, that he "is getting ready to go to work at four in the morning, even after a night of drinking." The faith that Trump was self-made was also embedded in her husband's affection for him. "A lot of people don't have that drive to work around here," she said. "And they live poor for the rest of their lives. I think they could make it if they wanted to, but there's a lot of laziness, a lot of scamming the system.

"Growing up in a small town, you know who the workers are. People may think we are prejudiced, but I hate all the guys in their pajamas buying sodas, just walking up in their pajama bottoms," she said, speaking of men in and around Scranton who were presumably underemployed.

"They are living in the projects, and women selling drugs are driving nicer cars than we do," she said, acidly. "That's why everyone here is for Trump."

———

Trump's bootstrapping fans were not all the usual suspects like Cole's husband and his friends and coworkers, or Roecker and his. One of the surprises of 2020, for instance, was the rise of the Latino Exit from the Democratic Party (LEXIT) movement. "Me as a Latina at fifty-two years old, I am working on my third degree," said a LEXIT-teer in September 2020 at a conservative Christian event called the Back the Blue Prayer Circle. LEXIT literature had a propagandistic flavor, including a refrain on "how the poverty-stricken areas are under Democratic control, which tells people that they cannot rise above their circumstances." Those in LEXIT have not only absorbed the self-made corporate narrative, but they also welcomed Donald Trump as the patron saint of such success. Trump's Latino voters in Florida and Nevada and elsewhere, according to *The Atlantic*, didn't "worry too much about systemic racism, because they think individuals should be able to pull themselves up by their bootstraps and take personal responsibility for their circumstances."

Some of the most vocal Black Americans who supported Trump would also take part in the bootstrapping refrain of the value of individual accomplishment and the weakness of communities in need. Rob Smith, a Democrat and veteran who became a Trump supporter, for instance, tweeted before the election: "As a black man, democrats want me to be a victim. They call me an Uncle Tom for talking about problems in the black community. Democrats want me to find empowerment in victimhood." Indeed, this ostensibly empowered the "antivictim" stance laced through pieces in conservative publications during that time. One personal essay in the *Daily Signal* was exemplary of this, asserting "the choice to thrive still lies in the hands of those who suffer."

This attitude was also telegraphed through interviews and online commentary. Brandon Straka, a Trump supporter and purportedly a former liberal, proclaimed, "I am a gay man who, on my own, without any help, without any political contacts, any money, created a successful political movement on a side of the fence where it's claimed I'm not even welcome." His "transition into conservatism," he said, which led to his #WalkAway campaign to convince voters to "walk away" from the Democratic Party, was inspired by those he perceived to be blue state snowflakes. As Straka said, "There really is no such thing as victimhood unless you choose to be a victim."

Part of the project of connecting self-sufficiency to Trump and candidates like him wasn't just about the present and the future. Even classic writings from the nineteenth century might be appropriated to the end of arguing for gun rights, say. Right-wing biographer Timothy Sandefur, writing on the free-market conservative Cato Institute site, repurposed the famed nineteenth-century Black writer Frederick Douglass. In Sandefur's telling, Douglass was not a visionary abolitionist with mixed feelings about self-invention but rather a "classical" libertarian who "believed quite clearly that the individual is the sole bearer of rights, and that the government exists to protect those rights." (Douglass did praise a "good revolver"—probably because he was once hunted as a fugitive slave.) Then in his book *Frederick Douglass: Self-Made Man*, Sandefur argues that Douglass is most importantly "a radical for individualism" and unconcerned about "the interests of the collective."

The version of Douglass appropriated by the right, as the historian David Blight writes, tends to circumvent his abolitionism. In 2020, I reread Douglass's classic lecture "Self-Made Man." Douglass was quite ambivalent toward the title concept. Douglass personally had a strong claim to the appellation "self-made"—he was born into slavery in Maryland, taught himself to read and write, loved revolutionary speeches, escaped from enslavement with the assistance of a free young Black woman, and became one of the most renowned writers of his day. Nevertheless, in lectures Douglass said that there were "no such men as self-made men" and "individual" independence "can never exist." And in his private life, as Blight writes, Douglass echoed this posture as "he easily admitted his reliance on friends and associates."

———

By later 2021, Roecker's attitudes had grown more fixed. The way the Democrats were responding to COVID was still a major issue in his farm community. "I personally didn't know anyone who has gotten really sick or died" from the virus, he said, though someone in his town would eventually pass away as a result of infection, and the nephew of a friend would require hospitalization. He and his farmer neighbors still felt overlooked, living in "flyover" country, as he said. "People are frustrated—that's what's bringing this on: everyone feels the politicians are lining their own pockets and taking care of themselves," he said. Roecker was also now using the first-person plural, as if he were speaking for the multitudes, as when he said, "We feel that the Democrats are taking away our individuality."

He seemed to be the human embodiment of individualism. We both lived in a country marked by its extreme divide between the singular and collective, at a time when these apparently opposing worldviews were colliding and rearranging in uncanny new ways.

7

Zen Incorporated

————————————

THE GROUP RELAXED on the chairs where they were seated. Then the meditation began, led by an instructor. She asked each member to note their body's physical presence in space. "Feel the sensation of yourself sitting in your seat," she intoned. I've got to get everyone out of the workday mode, the instructor would think to herself. "We are here and present," she said. The group was told to focus on their breath as it went in and out. The voice of their guide urged them to notice their thoughts and their senses and withhold judgment. The middle part of the hour they spent together was the teaching section, and in the last part the instructor answered the mindfulness students' questions.

This meditation class wasn't held in a blanched, airy yoga studio, though, and those who had gathered weren't sitting in lotus positions on bolsters. Rather, they were all listening through headphones from their cubicles at a giant health conglomerate. Their instructor, Suzanne Matthiessen, was a mindfulness coach tasked with helping them feel less nervous. Less anxiety, the company hoped, meant greater productivity, and more productivity meant profit. The year was 2014. The cover of *Time* had just announced "The Mindful Revolution." And while Matthiessen had spent most of her life studying mindfulness, she wasn't particularly happy with the news that it was going mainstream. "I just remember thinking, *Oh no*," she told me at her Hermitage, Pennsylvania, home.

Matthiessen understood that mindfulness, once the purview of spiritualists and seekers, was now a commodity. The trainings were "happening in corporate settings," she said. "And I was saying to 'em, Just let your thoughts float away like clouds." We were speaking over Zoom, and she appeared on the screen as an archetypal yogi, clad in a tie-dyed blue blouse and clattering silver jewelry, with her hennaed hair in a shaggy Stevie Nicks–style do.

When she worked in corporate mindfulness training, she led group sessions like the one described above, offering meditative respites to employees of several large health insurance companies. She offered three or four corporate classes a week, all an hour long.

At first, Matthiessen was loving her work. She felt she'd helped those she taught, who often complained of demanding or aggravating jobs, by ameliorating their pain and discomfort and better armoring them for their workdays with breathing practices and mental tricks. Then, over time, the doubt crept in that she was just allowing the companies to say they were doing *something*. One worker told Matthiessen how her manager screamed at her for taking an hour-long meditation and awareness class, the same class the company itself was paying for, presumably to eliminate stress. Why should people be asked to meditate when they are stressed by their business's weak institutional response to their professional problems?

Matthiessen worried that she was ultimately teaching relaxation techniques to make people accept aspects of their professional lives that they *should not*. While she advised workers to sit upright in a comfortable position and to notice their thoughts when they were in pain or dissatisfied, she was also questioning the practice: *How could they float away on a meditative cloud when they, say, weren't allowed to take breaks?*

"That's not reality!" she said. "And when workers were ill or going through a divorce, they still all had to be at work and high-performing and productive because their productivity was the key selling factor for why they were getting mindful classes to begin with."

Indeed, worker effectiveness has been the rationale behind a growing number of companies' offering mindfulness and meditation to employees on their premises. According to one *Harvard Business Review* piece published in March 2021, penned by five scholars, more than half of all large companies offer their employees some form of mindfulness training, which they define as a "broad set of practices and techniques focusing on increasing awareness of the here and now."

David Gelles, the *New York Times* business writer and a corporate-mindfulness guru, has urged companies to use these programs, arguing that mindfulness will make America not only better but also cheaper——that highly stressed employees cost $2,000 more per year in health care, while productivity nets $3,000 more per mindful employee. The chief executive of Salesforce placed meditation rooms throughout the company's San Francisco headquarters and had Buddhist monks stay at his house, because he had a "beginner's mind," the Zen Buddhist–endorsed mentality of openness and enthusiasm when studying a subject that had informed his management style. Corporate mindfulness is also one way for titans of business to spiritualize their single-minded efforts to aggregate capital. It was why billionaire Ray Dalio, founder of Bridgewater Associates, a massive hedge fund, zealously suggested Transcendental Meditation to hundreds of his employees, reimbursing them for half the cost of their $1,000 lessons. (He said of his passion, "I did it because it's the greatest gift I could give anyone—it brings about equanimity, creativity, and peace.")

But many of these companies offering corporate-sponsored meditation to their workers (they now included Apple, Google, and GlaxoSmithKline) were doing so in the belief that mindfulness was something to be acquired with the aim of greater worker efficiency, which in itself can be a cause of stress.

"You can't beat stress back like a lion tamer with all of these dynamics weighing on you," Matthiessen concluded. Sometimes, it was a simpler fix: you had to remove the source of your agitation. Stress in the workplace was like an overgrown garlic weed that sometimes needed to be dug up rather than just pruned.

Another problem of corporate mindfulness is that the healing rituals may transpire in the very spaces and among the very people that sometimes led to the workers' distress in the first place—according to a recent study, an "overwhelming 84% of respondents reported at least one workplace factor that negatively impacted their mental health." As the 2019 Great Jobs report recorded, more than half of Americans were profoundly unhappy in their jobs. The next year's Great Jobs report, conducted by the same prestigious foundations, was even darker: 40 percent of US workers said they had experienced worsening job quality since the start of the COVID-19 pandemic.

Back when she was a trainer, Matthiessen was sometimes privately messaged by her students with words to that effect. Most told her stress-relief practices did help calm them down in extreme moments, yet some also told her that the comfort did not last. They said they had little space in which to speak out about real struggles on the job, maybe because the Zen training they were getting was in fact *part* of their job (as a recent study of mindfulness programs by scholars at Brown University showed, *whom* we practice with is seemingly more important than how we practice).

Matthiessen recalled previous cultural crises that were difficult for such workers—the 2008 recession, for instance—when she would conduct meditation sessions from what was then her home office in Marin County, California, with her webcam picking up the Shoji screen—a Japanese aesthetic required by the company—in the background. "I couldn't go off script and couldn't tackle the things that were really bothering people," she said. "It was Mindfulness Lite." And for mental health issues, she said, it was "one size fits all."

Matthiessen felt that she "wasn't really serving" the folks she taught. It was "like a little burr" under her saddle.

At some point, and as a result of her concerns, Matthiessen left what she called "the modern mindfulness industry," which she feared reduced mindfulness to a pacifier for frustrated employees.

In the interim, corporate mindfulness had gone in dubious directions. In May 2021, Amazon's anxious warehouse workers were encouraged to step into a booth the size of a closet and use a computer to view mindful practices, which included screen-guided meditation videos and positive affirmations. The meditation booths were called AmaZen. That the booths resembled toilet stalls in size and shape was ironic, because workers at the company had said they had to fight for bathroom breaks. Others noted that the booths were akin to panic rooms or even "a coffin," as one observer put it. These mindfulness "practice rooms" stood within warehouses, and employees were told to lock themselves into the "individual interactive kiosks" during exhausting work shifts.

AmaZen could be seen as a solution that pressured individual workers to be the ones solving side effects of the overwork demanded of them. This is not the ancient wisdom of Buddhism, which is in fact a religion centered on not striving but being more awake and compassionate.

Matthiessen shuddered at programs like AmaZen, tweeting to me, "This is horrific—AND it's NOT how to practice #mindfulness!"

When I last spoke to her, she told me how she still believed the practices had a place at work, as long as they were not stripped of their ethical and spiritual components, so she had, in her words, started "creating tools" to better deploy mindfulness at businesses.

=====

Corporate mindfulness is not only practiced within cubicle culture. It has also become a massive consumer industry. The latter can be seen as a stand-in for other forms of care we should be receiving societally. As of late, it has been popularized through a variety of products: mindfulness is a business that rakes in billions (the mindful meditation app market alone was valued at $1.1 billion in 2021). Tens of thousands of books are being marketed to address mindfulness, including even volumes on mindful dog training (I confess I simply fed my own dog kibble and called it a day). You can make sandwiches with Earth Balance Original Mindful Dressing & Sandwich Spread. There are magazines, necklaces, mindful bubble cleanse and body scrub, but also, more importantly, investments in mindfulness training, not just in the kind of corporations I have mentioned but also in the US military and the police, with programs like Mindful Badge.

In other words, the tech ownership class has not only used mindfulness in their workplaces but profited by selling mindfulness: the ten largest English-language mental-wellness apps saw a combined ten million downloads for the month of April 2020, with nearly four million downloads of the app Calm. (It joined meditation apps such as Insight Timer, Headspace, and Buddhify.) Some of the apps were offered free to frontline workers but no one else.

Calm costs $69.99 annually, or for $399.99 there is a Calm subscription "forever." Headspace costs $69.99 per year, with annual revenues that were estimated at $100 million in 2019.

All these mindfulness apps and goods and coaching could be construed as simple consumer pick-me-ups, but they are also, at least partly, strategies to keep us trapped by self-reliance myths.

What people were using on their phones was an ersatz version of something whose origins that went back thousands of years to a variety of religious and secular traditions, like the Vedas from BCE. This includes the Buddhist concept of the "wheel of life," where greed, ill will, delusion, and the hungry ghosts of the dead threaten to take down the happy privileged gods.

While the American counterculture, in another cooptation of cool, embraced Eastern religions like Buddhism in the 1960s, the version of Zen that is most pertinent to today's office-block mindfulness kicked off the mid-1990s. That was when Jon Kabat-Zinn promoted a secular, Westernized mindfulness contained in two best-selling books and DVD sets of his meditation programs. While Kabat-Zinn was committed and well informed, the 1990s version of mindfulness that sprung up around him was sometimes closer to a T-shirt slogan—inviting participants to "DARE TO BE PRESENT," or, if you prefer, the more jarring "IT'S ALWAYS NOW"—than a rigorous practice.

This mindfulness was not just Buddhistic. It also absorbed elements of positive psychology, especially after the American Psychological Association's then-president Martin Seligman became a positive psychology booster, writing in the late 1990s that he was seeking to "change the focus of psychology from a preoccupation only with repairing the worst things in life to also building positive qualities." The positive psychologists had their work cut out for them. The likelihood of feeling happiness is not terribly high. If you look at studies measuring Americans' satisfaction you may discover, as the researchers of the National Opinion Research Center at the University of Chicago did in 2020, that only 14 percent of US adults report being *very* happy. According to the broader General Social Survey, with data on American attitudes since 1972, only 29 percent of Americans have ever called themselves joyful.

Over time, mindfulness techniques were presented as at least a partial answer to this unhappiness, especially in the workplace. And in time, the people who became most interested in mindfulness as a way to make themselves and others more content changed as well. Jaime Kucinskas, a professor of sociology at Hamilton College in Clinton, New York, has argued that mindfulness now heavily reflects the needs of elites.

By the aughts, the positive effects on the "mental capital" of Zen incorporated were blatantly white collar and celebrated in the business press. *Wired* magazine exalted mindfulness for "unlocking productivity." Over at Google, in 2007 its evangelist was Chade-Meng Tan, an early engineer at the company. The program was called "Search Inside Yourself," as if people's inner mental lives were actually search engines as well. It was an "internal course" for Google employees, as the website for the Search Inside Yourself Leadership Institute (SIYLI) tells us. Today, SIYLI, established in 2012 as an independent nonprofit organization, does mindfulness workshops the world over, as well as for companies like Procter & Gamble and Deloitte.

At the same time, academic studies praising the effectiveness of mindfulness training started to proliferate among . . . management scholars. (Perhaps it's not for nothing that sociologist C. Wright Mills writes to the effect that what was in the mid-twentieth century a new management science was just "an elaborate manipulation of workers.") To give you a sense of the incredible scale, one *Journal of Management* essay integrated the research evidence of some *four thousand-plus* scientific papers on mindfulness. Many were glorifying the practice. Or as one piece in the academic literature on the subject cheered, "Will mindfulness be another passing management fad? In a word, no!"

To help separate the wheat of valuable meditative practice from the chaff of New Age individualism, I met with Ron Purser, the author of the book *McMindfulness: How Mindfulness Became the New Capitalist Spirituality*. Purser is an unlikely professor of management at San Francisco State University who is particularly hard on Zen incorporated. An Erin Brockovich–era Albert Finney doppelgänger, he was a unionized electrician before he turned to academia, and Purser was particularly alert to social falsehoods.

"I knew how it felt to be second class," Purser said of how he had perceived his background of origin and his earlier days as an angry young electrician. It was back then that he started practicing Buddhism, however, and that was what saved him.

Now Purser was angry again, because the mindfulness that set him free had been co-opted, in his view, by the very corporate types he despised. "Mindfulness is weaponized," said Purser, with the end goal of "trying to yoke our mental lives to the company." To challenge what he saw as a scourge, he has a new podcast (that modern building block of resistance) called (chef's kiss) *The Mindful Cranks*.

In one of our conversations at the Rubin Museum of Art in Manhattan, in both the Tibetan Buddhist Shrine Room, where tourists bowed down before a golden altar, and on to the surrounding galleries, we were fittingly inundated with a visual culture that sought to reinterpret Buddhism. I asked Purser about an article he had written with David Loy that attacked mindfulness training in the corporate context. The two write that it "has wide appeal because it has become a trendy method for subduing employee unrest, promoting a tacit acceptance of the status quo, and as an instrumental tool for keeping attention focused on institutional goals."

The practice didn't tend to alter the indignities the workers experienced either, he said. The shallower parts of the mindfulness regimen—like the activity of chomping carefully on a piece of dried fruit—had their limits if they didn't come accompanied by necessary structural change. "Chewing on a raisin really does not help against exploitation and underpayment," observes journalist Esma Linnemann, a reporter whom Purser quoted to me as we passed a conceptual piece that re-created details of a 2015 earthquake in Nepal.

As we strolled, Purser further laid out his perspective: corporate mindfulness, he said derisively, is "Buddhism repackaged as a form of self-help." In addition, mindfulness programs like the one taught to cops, the previously mentioned Mindful Badge, he called "socially regulatory." While they may offer some individuals a helpful means of self-management, they also signal a way in which safety officers, CEOs, or schools could shift the conversation away from structural violence or company policy. As he said this, emphatically, I edged up to a mandala that was surrounded by debris and strange found objects.

"The idea of corporate mindfulness is that it can relieve stress, because stress is caused when an individual is maladapted to his environment," he said. The truth is usually the reverse—that our work environments do not meet *us* as individuals.

It might be easy to see Purser as lashing out at harmless, perhaps self-indulgent people—the kind who like high-necked calico prairie dresses and the Goop site—and frivolous corporate practices.

But the value of Purser's critique extended beyond all that. While I had met him before the pandemic, in the years following, more and more people started to catch on to the ways in which mindfulness was being marshaled as a cover for bottom-line individualism. During the early COVID period, even the tech industry–cheerleading publication *TechCrunch* dubbed mindfulness apps "a little opportunistic—as if the companies are using the pandemic and, in particular, medical workers' struggles to boost their downloads." The writer notes that if "the companies truly cared" about COVID-19's role in "users' stress and anxiety, a better strategy may have been one that involved rolling out an entirely free collection."

Later, the findings of a research study by five professors (a few were management-studies boosters of corporate mindfulness) included the observation that mindfulness "programs don't always improve people's well-being or their job performance." The study in the *Journal of Applied Psychology* found that the ultimate result can be "greater self-control depletion." A companion piece by the same authors, published in 2021 in the *Harvard Business Review*, noted that for employees "whose roles require them to act inauthentically," say, waiters, or customer service reps, many of them essential workers, "becoming more mindful of their emotions in the moment can actually have a negative effect on their mental health." These programs would sometimes give working-class people no way forward, because the only way forward, really, would be to change the conditions of their jobs, not their "headspace."

Mindfulness training, in other words, revealed the bad faith of low-paying jobs and the performative labor they require and then did nothing to expiate this newly clarified discontent.

Purser and other scholars who expressed criticism of corporate mindfulness included other trendy emotionally oriented regimens in the workplace and schools in their critique. He also cast aspersions on what I thought of as the grit business, which sprouted up in part thanks to Angela Duckworth's best seller *Grit: The Power of Passion and Perseverance*. Grit was now a part of American classrooms, with "grit workshops" where children (often those without means) were taught "grit skills," sometimes even including a metaphoric "perseverance walk." They might also be filling in "grit coloring book" pages. Indeed, grit became a great honorific for much of the aughts, a psychological shorthand for surviving hard times, a hardiness in the face of adversity, crediting individual pluck.

Grit has been critiqued, though: more than five studies in peer-reviewed journals over a few years, in addition to many articles, have questioned the concept. A handful of educators have also noted a class bias in how and when terms like "grit" are used. As Raymond Arroyo, a former New York City schoolteacher, blogs on Medium, "What frustrates me is that the discussion around grit is always in reference to low-performing schools." These schools tend to be in the poorest regions in the country, as Arroyo, who knew the situation firsthand, points out—and the assumption is that his former students are not "successful because they don't work hard enough." As education scholar Jal Mehta wrote in *Education Week*, summarizing this analysis, "The most prominent critique is that an emphasis on grit is a way of 'blaming the victim'—rather than take up larger questions of social, economic, and racial justice, if only the most disadvantaged kids were a little 'grittier' they could make it in life."

There is also a growing—though much smaller—backlash to other terms that put the overcoming of adversity back on the individual's ability to persist or even find joy in their capacity for endurance: words like *resilience* and *gratitude*.

In the consumer sector, the term could now be applied to "resilience coaches," one of whom taught the Boom Bounce Wow Resilience Method. Or at a lower cost than a hired consultant, you could buy a resilience best seller or "resilience planner" bearing the legend, say, "Stay Resilient 2019." In other words, focusing on these traits tended to skew conversations away from equity and the failure to thrive by these lights implied a lack of character.

One day I went to a bookstore and found one of these monthly resilience journals I had seen online promising things like "5 minutes a day" toward "balance." As I paged through it, I thought about the way Zen incorporated was both a reaction to and a reflection of our bootstrapped society. And then I wondered why employees and students were the ones who must adapt to the stressed workplace or middle school rather than the other way around. Was there a way to practice mindfulness, even at work, in a way that didn't paper over structural problems and put them back on our shoulders?

Pursuing this question was how I found a variant on mindfulness, through Purser and some of his colleagues. It's a budding countermovement that is sometimes called "socially conscious mindfulness." According to Rhonda Magee, a law professor at the University of San Francisco, that term entails not just being more aware of our sensations and breath but also the idea that the obstacles to such ease and awareness might be political in nature. Toward this end, Magee directs those seeking enlightenment and relief to reflect on "microaggressions—to hold them with some objectivity and distance" and consider them with a degree of detachment, all without denying the pain they may cause and internalizing it. In YouTube videos, Magee suggested that we openly discuss this topic.

It was David Forbes who led me to Magee. Forbes described himself, in a high-pitched Bernie Sanders–like Brooklyn accent, as a "professor of contemplative education," but his more worldly title is associate professor in the school counseling program at Brooklyn College's School of Education. He suggested that I check out the New Mindful Deal. That was his own invention, geared toward institutions that are teaching mindfulness to teenagers, police officers, soldiers, and other stress-prone groups. Forbes reframed mindfulness as "an awareness of social relations." At one high school in Brooklyn, for instance, where Forbes said he worked with school counselors, he advised against considering rage to be students' personal problem. Instead, students and teachers might "consider grown-up, righteous anger against social injustices" to be legitimate. Then, if students are allowed to reconsider these strong feelings as legit, they are likelier, he believed, to channel it effectively rather than just letting it "float away" like a cloud.

"There are a lot of well-intentioned people who are pushing stress reduction and being in the present," he said. These adherents "don't question what success actually means and buy into hidden assumptions."

After speaking to Forbes and listening to Magee and generally trying to find alternatives to some of the more popular institutional New Age practices, I considered what a better rendition of these actions might look like.

This was what led to my dropping in on a virtual mindfulness class. There were few preexisting relationships among this group of mostly strangers. It was free and thirty minutes long. Most strikingly, these sessions were seemingly available twenty-four hours a day for anyone. During one of the sessions I attended, a gentle-voiced facilitator named Lori introduced herself. She then asked us to spend fifteen minutes in silent meditation. The group sat with their eyes closed. I was to immediately feel a kind of peace: the quiet anonymity and the public togetherness of the program were reassuring to me. In fact, I was so relaxed and also so tired by my long pandemic days that I started to fall asleep.

At the quarter-of-an-hour mark, the facilitator indicated that we should stretch, wiggle our fingers or toes. We were asked to share experiences (only if we felt comfortable!). The day's theme was "obstacles," presumably ones we had faced. We were prompted to reflect on a difficult situation that a mindfulness practice helped us overcome with grace. The facilitator mentioned a recent chronic medical condition that had just been diagnosed: before she had found her practice, she "would have made stories around this" (apparently not good ones) and it would have become overwhelming. Now she asks herself if she can transform her body's suffering in some way: taking better care of herself and working less helped her cope with pain, she averred. I was stirred by her and others' stories and by the fact that these groups existed at all, without cost, without the insistence that participants become more productive or better earners.

At the close of a session, and after a few others like it that I attended, I realized that for me, authentic "mindfulness" was aligned with what thinker Lynne Segal names in her book *Radical Happiness: Moments of Collective Joy*. What Segal lays out is a considered way of being: what she calls "radical happiness" is deeper and far more outer directed than the glee favored by the "happiness business." (As Segal writes, "official" happiness is a "dubious development" because it disavows the depression and anxiety.) This radical happiness also emerged in days of yore when people danced in the streets together, as recorded by Barbara Ehrenreich in her history of communal celebration, *Dancing in the Streets*. By celebrating in concert, people attained a new degree of pleasure and awareness. I was looking for perhaps yet another variation on this experience, "public happiness," that which the great Hannah Arendt writes was an exuberance achieved when we moved away from merely personal concerns and instead participated in public life, producing a more complex collective joy. I'd wager the most effective form of stress relief is solidarity. It's certainly better than meditating in your office in the hopes of quieting the fear that you'll lose your job next month.

III

BURDENS OF
THE AMERICAN DREAM

8

Go Fund Yourself

JAMES FAUNTLEROY DOES not like depending on what he considers charity, but he has had to for the last few years, on the site GoFundMe. "I'm reluctantly asking for assistance until I find something that can supplement my income," he writes. "This is a challenge, though, but at least it won't be made worse by me being out on the street." And then he made a request for $5,000, which ended with a sentence characteristic of his extremely good manners: "Thank you for reading and I appreciate you."

Fauntleroy had suffered from poor health from a young age and was now burdened with end-stage renal disease. Since he developed the disease more than six years ago, Fauntleroy, who is in his thirties, stopped cooking full-time professionally, though his condition at first allowed him to work sporadically. Now he cannot work at all.

There was much more to Fauntleroy than this crowdfunding campaign and his malady, and I got to know these facets through conversations with him over the years. Sometimes when I spoke to him, he'd just returned from the other space that he frequents, the dialysis clinic. At one point he showed me his small room, where he both slept and worked, lit by the sunny Florida sky and abutted by a small mat of grass outside his window.

We talked about the week's events: Fauntleroy read the news voraciously—he even consumed Noam Chomsky for fun. We also discussed politics. He said he desperately wanted to live in a country that functioned better for people like him. A case in point: he had liked school, but never attended college, because he was afraid of going into deeper debt. We talked about what he liked to cook for himself and his mother. Though he was often broke, Fauntleroy had saved and had recently bought fresh ingredients for shrimp scampi pasta, featuring a white wine garlic butter sauce and sautéed mushrooms. He made it for himself and his mother, to her delight.

A self-described "people person," he even talked with me about the ideas I was wrestling with. "I am of the opinion there is no self-made man or self-made woman," he told me. "You had help getting to where you needed to go with roads and bridges, paid for by taxation. That's why you can drive around or why your drinking water is clean and you have public schools."

Talking to him, with his wry and warm demeanor and baritone voice, it was easy to see how Fauntleroy had once worked as a salesman, selling sectional couches and dining room tables that looked like maple but were made of particle board. Fauntleroy's LinkedIn profile still lists that job, with a profile photo of him looking sharp, with cherubic face and physique, in a checkered dress shirt.

Yet now he is too ill—and the pandemic too omnipresent—for that work, especially given that the company that once employed him received one of the highest OSHA penalties in recent memory of any company in the country.

What was really added to his mental load, however, was not his malady but the limits on his disability monies. He said it meant that he can't earn over about $1,200 a month and still receive benefits.

In addition, his Social Security benefits are at their max of $1,300 per month, which he described as "starvation income," or as he has calculated, what comes out to $6.60 an hour. "To expect someone on disability to live on this is immoral and indecent," he told me. He struggles to cover the $65 monthly co-pays for his medical care. Further, he has many necessary small expenses that are hard to manage, like keeping up the car he and his mother own, which needs new power steering; the hose has a leak. "I don't have enough money to get it fixed—anything above $30 is something we can't pull off very quickly," he explained on a crowdfunding site.

Like millions of other people, Fauntleroy relied on these platforms. He also believed that these sites shouldn't have to exist and these needs were not necessarily supposed to be relieved by an online charity drive, rather than government aid. In fact, like me, he felt these sites were evidence that our country had left us to survive on our own, and many of us could not do so, which was why many are at least partly getting by on what they are able to collect from friends and even near-strangers on the internet. It's part of what I call the "dystopian social safety net."

The dystopian social safety net includes nonprofits throughout the country that are forced to open "warming centers" to prevent those experiencing homelessness from freezing on corners and at bus stops during cold snaps. For slightly more solvent unhoused people, there are safe parking initiatives, where designated lots double as housing for those living out of their cars; in Walmart parking lots and other commercial lots, spaces meant to be briefly used by vehicles loading up on dog chow have become ad hoc overnight residences. The dystopian social safety net also includes organizations like the Patient Advocate Foundation with its Co-Pay Relief program, which can step in if you receive surprising and wildly excessive medical bills you cannot pay. The "dental fairs," free two-day dental clinics, are geared to the many people who cannot afford tooth care (fewer than half of American dentists accept Medicaid). These are fairs without Ferris wheels and bumper cars—instead, two thousand people at, say, a fairground are there to get their teeth cleaned or fixed.

The dystopian social safety net included several Alabama public school teachers who donated rare and precious time off to a fellow teacher so that he could spend more time with his cancer-stricken sixteen-month-old daughter who was getting treatment at a hospital a hundred miles from home. It was also composed of that group of colleagues who'd told me how they combined their vacation days and gave them to a colleague to fund a "paid maternity leave," because their employer had provided her with none. It included people like Keoni Ching from Vancouver, Washington, who was eight years old when he started to sell handmade key chains for $5 each on the site. Ching was selling these trinkets to help kids at his school, and his act was followed by positive press and requests for the chains from Alaska and Arizona. He ultimately raised $4,015.

The dystopian social safety net is what often patches over holes in our systems, but the organizations and so forth that compose the net are not always above reproach.

Take the drug rehab organizations like the one that offers one lucky caller (!) the free recovery help they need. An even more repellent example of this was when a South Dakota hockey team and mortgage lender in 2021 threw a promotional event called "Dash for Cash," where underpaid schoolteachers were encouraged to crawl and slither on a rink's ice grabbing $5,000 worth of dollar bills off the cold surface and stuffing them in their clothes during an intermission. This filmed occasion went viral.

And when the pandemic hit, Americans' dependence on the dystopian social safety net only grew. Families used GoFundMe to underwrite their food costs while they stayed at coronavirus isolation hotels or to pay for Zoom funerals after COVID deaths. Others campaigned for essential workers, for beloved bookstores and restaurants set to close. A GoFundMe page raised over $50,000 for two Black FedEx drivers who claimed they were fired over a viral video confrontation with a customer in May 2020. "All we did was deliver his package; he was in the house at the time," said one of the deliverymen in Leesburg, Georgia.

The dystopian social safety net, a mesh of programs that could be necessary only in a failing society, is not usually acknowledged.

These stories exemplified the trouble with the dystopian social safety net, as it runs on news stories about such altruistic efforts that offer readers or viewers bursts of hope. Yet they also point to how tiny is the number of people who actually manage to get their needs met by these shadow social services.

There is no shame in our needing to use these sorts of Band-Aids. Fewer than half of American adults—just 47 percent—say that they have enough emergency funds to cover three months of expenses, according to a survey conducted in 2020 by the Pew Research Center. There is shame, however, on the part of our government that we are required to put on codified theatrical performances on hockey rinks or on crowdfunding sites if we are to get what we need.

Fauntleroy was one of the army of people mounting these performances of self on GoFundMe as if it were a state program; it was not just for his medical care or to fix the car but also to pay for the small apartment he and his mother share. The need is so great, in fact, that GoFundMe and disability are not enough, and each month to cover their rent he had to take out a payday loan from Amscot, a chain that calls itself "the money superstore" and where the service fee is 10 percent of the amount advanced plus a "verification" fee. That a payday loan shop was the most convenient shopping excursion in his roughneck East Millenia neighborhood was no coincidence. Fauntleroy had gone there once a week for seven years—the employees at this predatory loan outlet know him by his first name and even how much money he needs.

Many depend on the dystopian social safety net not only because of the failure of government support but also because of the failure of mechanisms that allow people to access federal or state support. This battle to obtain social benefits has a name: the administrative burden. While there are few studies on how much time and effort people put in to get the services they need, scholars such as Georgetown University public policy professor Pamela Herd, coauthor of the book *Administrative Burden: Policymaking by Other Means*, think a good measure of the administrative burden is the "take-up rate"—an indication of how many people actually make use of a safety net program, compared with the number who are eligible.

These rates are the result of the swirls of red tape Fauntleroy and others have encountered when they try to access the simple government benefits they are eligible for. A low take-up rate strongly suggests that getting aid can be unduly burdensome. For SNAP, the national take-up rate is about 82 percent, but it can be as low as 52 percent in some states.

What these low take-up rates mean is that in certain states eligible populations are not accessing major programs. This includes Medicaid, whose take-up rate varies widely by state—from roughly 90 percent to less than 50 percent. Bureaucratic hurdles can take many forms: in Tennessee, for instance, an investigation by the *Tennessean* newspaper found that in recent years, 220,000 children had been kicked out of the state's Medicaid program because of clerical errors—"In a system dependent on hard-copy forms and postal mail."

Still others like Fauntleroy struggle to get their payments, all against the backdrop of there not being enough to survive on to begin with. Fauntleroy also can't get rental assistance due to another arcane ineligibility.

Herd and others argue that administrative burdens like these are no accident: these systems are often *meant* to be difficult to maneuver—because their recipients are viewed skeptically by many Americans. Contrast Social Security's take-up rate (almost all older Americans, an estimated 97 percent, according to the Center on Budget and Policy Priorities) with that for many other programs. In moments of frankness, politicians will sometimes admit the truth.

"Having studied how [the unemployment system] was internally constructed," Governor Ron DeSantis, the Florida Republican, told a CBS News affiliate in August 2020, "I think the goal was for whoever designed it was, 'Let's put as many kind of pointless roadblocks along the way,' so people just say, 'Oh, the hell with it. I'm not going to do that.'"

The widespread existence of the administrative burden is part of why private and nonprofit networks enter the fray as semi-saviors, a valuable if flawed model for transformation. Improvisations like homeless housing in parking lots are a grim necessity and a response to the American template of self-sufficiency. Large societies *require* large and functioning public institutions to coordinate and manage the needs and struggles of their millions of inhabitants, rather than just relying on a baroque lattice of not-for-profits that are themselves constantly being asked to be more self-reliant.

To make matters worse, in 2018 the Trump administration allowed states to require that Medicaid recipients spend a certain number of hours working or seeking work to continue to receive benefits. The results were catastrophic. Instead of inspiring financially stressed people to find jobs, the new policies simply introduced more convolutions. Thousands lost coverage, despite meeting the work requirements, and a Harvard study found no significant change to employment status. While the Affordable Care Act of the Obama administration had led to gains in insurance coverage overall, health-care costs continued to plague many Americans, and in 2017, eleven million Americans said they had had "catastrophic medical expenses." GoFundMe, created in 2010 with lighter campaigns in mind, has become a major social service provider: it has gotten to the point where even the company's founder complained publicly about GoFundMe's new role: Robert Solomon, the CEO of GoFundMe, told CBS in 2019, "We weren't ever set up to be a health-care company, and we still are not."

The public school system, Medicaid, SNAP, and . . . GoFundMe? It can seem as if these are the branches of today's American welfare system. Fauntleroy's was just one of the roughly 250,000 GoFundMe campaigns related to medical care in 2018, and as of September 2020 more than $625 million had been raised for these COVID-19 relief campaigns. Even before the pandemic, one-third of all GoFundMe campaigns—approximately $7 million in 2018—were for unpaid medical expenses, according to the company's CEO. According to NORC, at the University of Chicago, more than twelve million Americans have begun crowdfunding campaigns to help someone outside of their immediate family afford their medical bills, and eight million have begun such campaigns for themselves or a member of their household.

In addition, for the many Americans like Fauntleroy who also need an organ transplant, the cost is too high. He cited a number in the five figures, but others have noted that averages are in the range of $400,000 to $1.4 million.

These needs are so pervasive that GoFundMe even offered tips for how to squeeze what you can out of strangers and friends, advising "A great campaign story." That story, the company wrote, "will outline your cause clearly, in a way that is engaging to read . . . all while speaking from the heart." GoFundMe's American dystopian mode includes commodifying our suffering for entertainment; we are forced into social performance, like actors, learning lines that are beggarly, with plaintive intonations, a veritable bazaar of pain.

Another striking example was Annie Hanshew, who raised money on GoFundMe for school lunches. (Youth hunger is another thing addressed pretty frequently by the dystopian social safety net. Even before the pandemic, 640 colleges and universities operated food pantries on campus. In 2016, a survey of almost 3,800 students in twelve states found that nearly half of them were food insecure.)

Hanshew had wanted things to change in her hometown of Helena, Montana—she even ran for the local school board. Her activities came to a head when she found out in 2018 that a collection agency planned to squeeze more than $100,000 from families who hadn't paid for school meals. Hanshew then reached for the tool of crowd-funding, one many Americans have resorted to. She soon discovered that school district officials hadn't tracked which student debtors actually qualified for either free or reduced lunch: she knew that this move would further burden area low-income families.

She then went on GoFundMe and saw that citizens in Seattle and Fargo were using the site to raise money for school lunches in their community. These efforts were designed to keep kids with unpaid meal accounts from getting "lunch shamed," as some are, with phrases like "I need lunch money!" stamped on their arms by workers at public schools, a practice a number of children endured.

In the past, a local activist like Hanshew might have simply called their school board or complained to an elected official, which she did, yet school boards themselves weren't always advocates for their students, and their members could spread the stigma around to the smallest citizens without recourse. A suburban Milwaukee school district's board members were reported to have tried to opt out of a federally funded free meal program: one member declared such offerings led families to "become spoiled." Another school administrator worried out loud about free lunch becoming a "slow addiction," a corrosive mentality that stands in the way of people getting the help they actually need.

Which is why, today, the dark futures promised by sci-fi films like *Mad Max* or *Elysium* or *Children of Men* or *The Hunger Games* are upon us, where the wealthy live in luxurious palaces or opulent space stations and the rest dwell in a world of ordeals and forms and social media charity campaigns to survive.

As Hanshew told me of her specific school lunch battle and GoFundMe campaign: "In a perfect world, kids would just get meals without means testing. But I am so used to this reality that I can't imagine change."

The dystopian social safety net that Fauntleroy and many others depend on is at least partly the handiwork of American legislators and the programs they created, or did not create. It seems as if we are stuffing the holes in the hull that is America with rags, but the view from steerage is that this ship is going down.

This was the setting in which Fauntleroy became a passionate supporter of Senator Bernie Sanders, from even before the 2016 election. Black and queer, he saw Sanders as the candidate who understood being marginalized or different, as few other politicians could. His greatest hope was that Medicare for All, one of Sanders's key proposals, would pass and then many of his efforts to raise money, he imagined, would no longer need to exist and his life with renal disease would then be a whole lot more manageable. In this dream world, he might even receive a new kidney.

Fauntleroy's zeal for the avuncular socialist has lasted through Sanders's 2020 loss and beyond. Talking with Fauntleroy, I, too, got a glimpse of the fantasy future he had been living for, one where Sanders's universal health care would then be passed into law. GoFundMe could go back to doing what it was intended for—paying for film projects and weddings—rather than covering so much of America's medical costs.

By late 2020, he was crushed by the end of the Sanders campaign, and in the immediate aftermath, in early 2021, he needed money again badly. He was considering selling some of his things on eBay: dress shirts, polos, suits, and blazers, and maybe eventually putting up another GoFundMe.

While Sanders had lost, Fauntleroy was still primarily receiving emotional support and even sometimes physical support from other Sanders supporters and enthusiasts he often barely knew. They checked in on him online and direct messaged him. "I can't go out and see them, of course," he said. A follower on Twitter who was a political fellow traveler doing some Florida tourism picked him up at his apartment and went with him to a Starbucks for two hours, accompanied by his wife. It sometimes felt to him as if the Sanders campaign itself had become part of the dystopian social safety net. He was relieved of the loneliness only when in the winter before the pandemic, he attended a Sanders watch party at a local bar for the second Democratic debate and someone from the campaign helped him get there. He worried about the effect of realist Democrats whose supposedly rationalist perspective excluded his and others' needs. He found theirs to be an attitude of nihilistic realism.

He understood that what props us up is part of what makes us human, and the available resources are not enough for him. In October, he tweeted out almost poetic evidence of scarcity: "I've only seen more than 10 stars in the night sky only once in my life."

He'd mostly stayed in his home throughout the pandemic to avoid getting sick. When he went outside, it was often to, say, mow the lawn. He thought that during this crisis he was seeing something grim, but that it would ultimately lead to change—that it would narrow the difference between lower-class and middle-income people. "Being of lower income you don't go out anyway, you don't do things anyway—it's the same exact things as it's always been," he said. "I am always on a lockdown economically," is how he put it. The majority of Americans, he continued, "are going through what we go through, they are in our shoes, they have unemployment, they have to stay home."

Fauntleroy's boyfriend, a photographer who lived in North Carolina, was also under lockdown. Because his partner was also in the red, Fauntleroy felt he couldn't ask him to relocate to another state and move in with him and his mother. They were divided by their income levels and how little this country had done to relieve them. He said, "If something goes wrong, we are both out on the street." He told me he was thinking a lot during this time about "the gigantic disconnect between people like me and those in power."

What could be done to shrink this gap? One way would be to emphasize the experience of those enduring poverty through the use of roadside signs, public service announcements, and political speeches, so they are not forgotten.

Another would be more canny and imaginative organizations representing their interests. A good and relatively recent example of this has been the Poor People's Campaign, led by Reverend Dr. William J. Barber II, a movement that sought to represent 140 million poor people and included holding events at the West Virginia Capitol to exert pressure on Senator Joe Manchin to support the bill for crucial social services.

In addition, we can learn from the pandemic realities how to adjust and ease the administrative burden. COVID had, paradoxically, helped to cut the red tape for welfare benefits. For instance, it was announced in March 2020 that people applying for SNAP benefits no longer had to undergo in-person interviews. In such interviews, applicants are typically quizzed about such matters as their income, the size of their credit card debts, and how long they have lived in their current residence—when most of that information is readily available without an interview. SNAP interviews are basically pointless—and can include, according to people who have experienced them, prying about such issues as the cost of children's clothes and toys. For a while, they had been put on pause.

Similarly, as of March 2020, people enrolled in Medicaid no longer had to undergo "recertification"—a regular review of their records to ensure that they still qualified. The process involves substantial paperwork for families and is often conducted through the mail. Partly for that reason—as well as increased demand—Medicaid enrollment by the end of November was up 20 percent since the beginning of the pandemic.

The policy shift also had a striking effect on "churn": the cycling of people off and on Medicaid: normally, in one year, 11 percent of children who are enrolled and 12 percent of adults would be disenrolled and then subsequently reenrolled. That pointless pattern essentially came to an end.

One way to advance from where we are, and away from dependence on GoFundMe, is to spread the recognition that "the administrative burden" represents a systemic injustice and get this message to those who actually suffer from it: the recognition that frustrating applicants can sometimes be baked into these systems would alleviate some of the self-blame and perhaps channel it toward efforts to change the system.

Policymakers should also, where feasible, make permanent the pandemic-era simplifications of programs like SNAP and Medicaid. Observers expect that when the pandemic emergency is deemed over, this could cost millions of people their health insurance. Some form of recertification probably must exist, but a more streamlined version that causes fewer people to lose coverage because of missing paperwork is not only advisable but necessary.

Still, as we wait for such changes in systems like Medicaid and SNAP to be made permanent, we have the dystopian social safety net.

Fauntleroy himself tweeted about the cruel paradoxes he and others are caught in, to increasingly enthusiastic followers in early 2022. Yet he also held out some subdued hope for the future. As he said, "Living in poverty means I have been tempered by fire."

9

Mothers' Revolution

In those years, people will say we lost track of the
meaning of we, of you we found ourselves reduced to I.

—ADRIENNE RICH, "IN THOSE YEARS"

MONICA SCOTT'S ELDEST, Da'Quaylon, is only fourteen years old,
but during the pandemic he had to become a child laborer and take
a job—unpaid—as a babysitter. For many months, he looked after
his brothers, aged seven and nine, day and night, while his mother
worked, often seven days a week, to support the family. The family lived in Lakeland, Florida, the westernmost city of Polk County,
which, at the time I got to know Scott, had one of the highest rates
of COVID in the state. The case numbers were so high that even
the smallest liberties and pleasures had eroded for the kids: the little
ones were not even allowed to walk to the nearby ice cream truck by
themselves anymore.

After the pandemic struck, the Boys & Girls Clubs of Polk
County, which once offered after-school care, shut down. As a result,

Scott had to leave her sons on their own when she went to work at McDonald's. She was paid $9.25 an hour, even though she had been there for four years, with only a tiny hazard bonus. To make ends meet, she took on a second job at Amazon sorting boxes at the airport from one thirty in the afternoon to ten at night. Despite all of her hard work, she and her children couldn't afford a permanent home: the family lived in a motel. While she was still searching for day care or babysitting, she worried how "trustworthy" any cut-rate childcare option would be. "It's a lose-lose situation," she said.

Also, when the boys' school went remote, she didn't own a computer for the two of them, who were now locked in at home with their elder brother, which made their schooling even more tangled and desultory. The "little ones" found it hard to cope, and when she got home exhausted, she had neither the training nor the bandwidth to handle her children's educational woes. Her then fourteen-year-old would get his homework done, but the nine- and seven-year-old wouldn't do their assignments, and at the end of the school year she was grateful that they had passed their classes. And all the time she was working, working, working. She said, "It was tragic for me, I never worried more than I did at the time when they were homeschooled—there were no after-school programs where they can get help."

Scott was far from alone. She was one of millions of mothers in this country for whom childcare is a bewildering crisis. Scott's distress didn't start in 2020, but the virus had made her quiet challenges more extreme. It wasn't her fault. For decades, the United States had absolutely overlooked day care as a matter of course, viewing children's care as a private need, not the job of the state, something left undiscussed. Scott's and other mothers' struggles were thus silenced. As one expert told me, these mothers were trying to get their family's needs met not from a failing system but from a failing *nonsystem*.

In Scott's case, the childcare nonsystem meant that while she applied for day-care vouchers, she had to earn less than a certain income threshold. "I went over the mark, but we are still paying the rent, paying the bills—when bills are due you have to pay them," she said. To top it off, the only rent the family of four could afford was for lodgings at a motel, where Scott had to be cagey with the manager, since no children were officially allowed there. Scott had planned a different path, getting a cosmetology license paid for by Federal Pell Grants. Then she got pregnant in her teens and had to decide whether she would keep trying to get jobs as a hairdresser and the like or dedicate herself to her baby. Her great-grandmother had always been on her mind, the woman who raised her when it wasn't "peaches and cream," when her mom, she said, left her when she was a year old. (Her father had passed "when I was in my mother's stomach.") Later, there was "oppression and depression" to deal with. Her mother's negative example, she said, "kept me from being the mother who was like that. I always told myself, Lord, when I have kids, I am going to be here for them. I feel this in my heart."

Her children's fathers have "checked out," she said: one was deported, and she did get child support from another but never saw him. She also tried other things, long before the pandemic, but the childcare part of it made these jobs impossible: a warehouse job where "the money was awesome" but where she would have to be away from her kids when they were really little during the time they were home from school. Moms like Scott found a way through informal networks as well, although it was far harder than for more middle-class mothers. In August 2020, month five of the pandemic, Scott found someone in the neighborhood she could trust, who attended the same ministry she did, to look after her kids; the woman had her own infant daughter. Who knew how long that would last—many of these sorts of arrangements dissolved around her, as they did for so many parents seeking to create a day-care net out of unraveled string.

Parents like Scott—mothers in particular—have long been on the losing end of the American obsession with independence. Having children in this country can, in fact, be the fastest way to having the American Dream disproved. How are we to succeed on our own with young children and without the promise of community or assistance?

This was further dramatized during 2020. By October of that year, two million women had fallen out of the workforce, sending us back to 1988 levels, back to the age when being a "workingwoman" was so notable that there was even a magazine of that name featuring cover girls in navy shoulder-pad-heavy blazers. By some estimates this was four times more than the number of men that had fallen out of the American labor force each month for six months straight. Unemployment filings further revealed how we were capsizing, until the US Census Bureau revealed that in January 2021 ten million mothers with school-age kids were not working, an increase of 1.4 million compared with the previous year. (The way in which maternal workers disappeared from the workforce in this time reminded me of a suspicion I had long had about the self-made story—the mother is often erased from it, as bootstrapping insists that a person flourish alone, sometimes as if they'd never had a mother in the first place.) In 2020, nearly two in five caregivers surveyed by the University of Oregon were concerned that they wouldn't have enough food for their families in the next month.

These mothers were engaging in a fundamental activity, what theorists have called "social reproduction." That's because parents create children who will become adults who earn wages, engaging in the society-building endeavor of making little citizens. The paradox is that while this is an essential activity, one doesn't get into the self-made pantheon by parenting.

These truisms were wildly exaggerated during the pandemic. The mothers who managed to keep their jobs (according to the economist Katica Roy, it will take years for women's employment to climb back to its 2019 level) were often educating their children at home, at least part of the time, during "hybrid schooling" or taking care of preschool-age children and infants while still doing their work for pay, their days set up like obstacle courses. Data from the Rapid Assessment of Pandemic Impact on Development—Early Childhood Project at the University of Oregon told us just how much the pandemic day-care crisis had strained families, specifically those with a low income. Parents' search for day care alone was at once gratingly banal and tense and full of unknowns.

This included mothers I spoke to like Dasja Reed, then thirty and living New Orleans, who had failed for a while to find an open and affordable center during the pandemic. She told me that, as a single mom with a two-year-old son, Jarret, she couldn't go back to her job at the New Orleans Recreation Development Commission and had to take unpaid time off. That lack of day care ultimately put her and her son in jeopardy, an eventuality that lined up with findings from an August 2020 Center for American Progress report, that "licensed childcare is more than three times as scarce for children ages 0 to 2 than it is for those ages 3 to 5."

"For many, many families, prepandemic day-care emergencies," as one organizer and mother put it, "are now full-blown crises."

After talking to many mothers during the pandemic, I wondered if the obsession with our taking responsibility entirely for ourselves—the self-made myth—was not coincidentally antimaternal but purposely so, perhaps even adversarial to our biological origins. As one writer puts it, "Babies are radically dependent. We all are, really, but some life stages lend themselves better to maintaining the illusion of total autonomy." And thus babies' parents or guardians—often their mothers—are the ones most closely engaged with and acknowledging the radical dependence that all people have on one another. They are also often the ones blamed for and saddled with babies' and children's fragilities, as well as the frailty within the whole apparatus of American youth. As the writer Jacqueline Rose argues in her book *Mothers*, motherhood is "the ultimate scapegoat for our personal and political failings, for everything that is wrong with the world, which it becomes the task—unrealizable of course—of mothers to repair."

This led me to another thought. Where was the anger at what

had been thrust on us? Where was the parents' revolution? Or even the parents' collectives? Parents may like such a fix in theory, but it's hard to pull off. It's a "great idea to start a parenting movement," Kim Kuban of Mount Vernon, Indiana, a teacher who said she now cared for her grandchildren, writes to me, but "who will be the movers and shakers? Parents are already worked beyond what is healthy." When she was younger, she said, she "had zero energy to reach out. Everyone in my circle was in the same boat, so I felt isolated from the 'outside' world of possible help."

And the pandemic also brought another sapping element of mothers' lives to light: the longtime gender inequities within American homes. In locked-down households, childcare fell greatly upon the women (I focused on households where there were two parents who identified as male and female respectively, when I looked at this phenomenon). One couple, adjunct professors in the California State University system, once were planning to send their toddler to a subsidized on-campus day-care provider, but when the campus closed they scrambled, struggling to send their seven-year-old and two-year-old to a camp and day care, respectively. The total cost was $1,600 a month, almost as much as their rent, eating up much of what the two earned. The couple was prepped for herculean course loads during a pandemic semester—eight courses for the husband and six for the wife. As the mother said, even if Dad was in charge: "I'm at the computer, and they're over there: 'Mommy, mommy, mommy.' . . . It's hard for me to push my kids away. . . . And I'm like, 'Well, dude [her husband], you need to figure out how to interact with them better so that they can feel safe to come to you for emotional stuff.'"

She wasn't alone. Women were not only much more likely than men to lose their jobs or be furloughed, but they were also far more likely than a male parent to be tasked with managing their children's schooling during the pandemic. I knew that this experience of highly gendered generosity around care was sometimes called "asymmetrical giving." During COVID, mothers could seem like they were playing out the also very gendered story in the children's book *The Giving Tree*, where the tree, a mother stand-in, gives and gives everything she has to the child until there is nothing left. (How I loathed that book, even when I was a young child myself, and I hated it all the more with respect to my own life during the pandemic, cleaning up while the work UberConference played on a speakerphone.)

In other words, many female parents (and those who identify as female) were not given the space within their families to participate in the system of American success, however toxic it is. How were these mothers to manage a shellacked individualistic pose when pandemic parenting—breastfeeding and carrying around snotty tissues and pretzels in Ziploc bags that disintegrate in their purse, or leading their child through "gym" in a tiny green space nearby and then also in Zooms for meetings on the phone while they did so? How were moms to balance workplace budgets while cooking every meal, including that once kid-free meal, lunch, cutting carrots for dinner, and making sure the mathematical magic squares added up?

In contrast, fathers have been found by researchers to be more likely to maintain their personal and professional boundaries and tend to be more self-preserving professionally, and the pandemic further brought this to light. Working from home even increased their productivity: according to an article in the *Harvard Business Review*, "a recurring finding is that women are more likely to carry out more domestic responsibilities while working flexibly, whereas men are more likely to prioritize and expand their work spheres." This was "asymmetrical giving," where women contributed so much more than their male counterparts to their children and their elders, their colleagues and suffering from COVID family members, even their friends and their children's classmates.

The interlocking structures of motherhood and professional childcare have both been starved of resources for centuries. In the late nineteenth century, at a time when the carapace of our fragile childcare system was created, dependent populations like mothers and housewives were themselves subject to bias, as the scholars Nancy Fraser and Linda Gordon write, because they were excluded from wage labor. As I have mentioned previously, the supposedly "independent" populations—able-bodied white men, often husbands and fathers—paradoxically were intensely needy. They just economically leaned on the invisible labor of their wives and their servants and enslaved people, people who were cast as supposedly "dependent" on them.

In later eras, women who required childcare in order to go to work, particularly poorer women, were nevertheless stigmatized. Day care was intended from the start to be a weak system, as scholars of the history of childcare tell me, like a punishment for needing care because a woman didn't have a husband or some other means of support. Racism had also long played a role in the childcare system. Many of the working mothers early on were women of color, and some of the first day-care providers were enslaved people caring for white children. The disdain for day care as a system also was a disdain for dependence. The aura of dependence around infants and children, which also adhered to their female caregivers, is to this day why they are poorly paid and why childcare itself is stigmatized as if the mewl and cry and spit-up of our youngest were secrets that should be kept behind the closed doors of a home.

In the 1960s and '70s, these biases came into question in some arenas. This happened in policy circles and also among feminist thinkers like the conceptualist artist Mierle Laderman Ukeles. Ukeles, one of my favorite figures of this resistance, defined herself as a "pissed-off" mother, writing in 1969 a manifesto about how sick she was of her role's endless and wearying tasks. Through her "formal education in autonomy," Ukeles was particularly struck by how the burdensome caregiving of kids was akin to the labor of paid cleaners and sanitation workers, the latter earning low wages, "doing support work, to keep something else going, and not necessarily only themselves." Maintenance work was, like motherhood, Ukeles writes in a delightful comparison, "the sourball of every revolution."

Two years later, in 1971, President Richard Nixon lobbed another sourball at the revolution when he vetoed a major bill proposing national day care, saying that such a measure would "commit the vast moral authority of the National Government to the side of communal approaches to child rearing. . . ." In other words, Nixon derided universal childcare as a sort of proxy Red Menace. Sixteen years later, President George H.W. Bush in 1988 famously lauded the "thousand points of light"—American charities—when he accepted the Republican nomination and prescribed such altruism as the societal answer for needy families, rather than government social programs. The title of Marvin Olasky's 1992 book *The Tragedy of American Compassion*, which generations of politicians on the right have fanboy-ed over, said it all. Olasky wrote that in poverty relief for families, "The goal is to look within the family first; if the family cannot help, maybe an individual . . ." This argument tended to include an idea of a time in the past—in the eighteenth or nineteenth century, say—when sickness and economic suffering were combated by antique ladies-who-lunched, all without "meddlesome" government interference. Those with this outlook may still ask, *Why are there no workhouses?*

But it wasn't just Bush 41 who insisted on families' and parents' evincing an almost impossible level of financial independence from the government. It was President Bill Clinton who famously said he would put an "end to welfare as we know it" and proceeded to do just that, limiting families to five years or fewer of benefits, undoing Aid to Families with Dependent Children and replacing it with the more fragile sustenance of Temporary Assistance for Needy Families, a program that made it harder to access funds. This was a move that insisted that whole families, a number headed by mothers, bootstrap themselves to financial security. It would also show how unfeasible that was: in the intervening years, families living in extreme poverty increased by 50 percent. These strict welfare laws threw their former recipients off the rolls, and when they couldn't find employment or jobs that paid enough, things went from bad to worse. (The moment I read about "workfare" in the newspaper in 1996, as an idealistic twenty-something, was the moment I stopped automatically trusting Democrats.)

More than a quarter of a century later, we still don't have national day care, adequate permanent support for poor families, or an answer to the question of how parents without resources could ever hope to pull themselves up by their bootstraps.

Embedded in this fallacy is this: the wealthier women who were said to "have it all" quite obviously couldn't actually *do* it all, so they usually hired others to do the invisible labor that keeps up their effortless appearances. As Allison Pugh, a sociologist of family life at the University of Virginia, said of this, "They have been using other women's paid labor—mostly women of color—to hide those conflicts from themselves." (I emphasize women and mothers and those who identify as women in this chapter as these are the people likeliest to do unpaid care work.)

What was missing, in part, for these women is that there was still little care infrastructure. The United States spends a lower percentage of its GDP for childcare than many industrialized countries, according to the Organisation for Economic Co-operation and Development. The caregiving of kids takes up 1.6 percent of the GDP of Sweden, whereas it's less than 0.5 percent of the domestic product of the United States, the sort of small budget that pays for, well, stacks of coloring books rather than actual day care.

All of which leaves parents to get by on their own, a version of parenthood that is an anomaly in the developed world.

There were parents in difficult monetary straits who were weighed down by demands of our country's individualistic notion of family-hood, like Scott. There were also the mothers who on the surface may have looked like they were doing just fine but had nonetheless been radicalized by how parenting didn't fit into their bootstrapping aspirations, especially during the pandemic.

Among their number were podcaster and editor Katherine Goldstein, who, in her description, started out as "this hard-charging, climbing-up-the-media-ladder person, achieving a lot of success in my later 20s and early 30s." Her family and schooling had taught her that "the harder you work, the luckier you get, that you will definitely succeed and be rewarded." In her early professional life and her earliest days as a parent as well, she loved the corporate feminist view of women's empowerment. She was, for instance, a superfan of Sheryl Sandberg, author of the 2013 best seller *Lean In*, and who was worth an estimated $1.8 billion dollars in 2020 (the only woman on Facebook's board of directors). She read Sandberg's book and was allured by its photo of the author smiling demurely, dressed in business casual. She watched Sandberg's TED talk on women's leadership, where the speaker opined on why there are so few women leaders while gesticulating with semaphore-like precision. (My favorite parody of a line from one of these ubiquitous talks appeared in an article by the scholar Jedediah Britton-Purdy: "'Hello, there is literally nothing we can do to change the course of this global death cult, thank you for coming to my TED talk.'")

In that talk, Sandberg recounted her experience "pitching a deal" in a private equity office where she realized that she may have been "the only woman to have pitched a deal in this office in a year," all to illustrate that only 15 percent of women make it to the "C-suite" and also, presumably, that Sandberg herself had made it. Or that Sandberg wrote that women hold themselves back "by lacking self-confidence, by not raising our hands, and by pulling back when we should be leaning in." Goldstein adored this I-can-do-anything archetype so much that she joined something called a "Lean In Circle." There were many around the country then, small groups of women, all connected by the book. Goldstein had bought into the corporate-feminist-slash-mother narrative, affected, too, by the girlboss "rich fiction" we read about earlier.

The Lean In Circle was intended, Goldstein said, to raise the corporate consciousness "with supercorporate bullet points women were supposed to follow." The women would meet and discuss their jobs. It was "very cheesy," Goldstein said, but she still was engaged by the experience and liked the women she had met through it, one of the first times she "had a dedicated group of women" to discuss professional things with. "They were all white-collar professionals and college educated, but not necessarily wealthy people" in their late twenties and early thirties, she said. The circle seemed to give her a sense of validity that she had been looking for: "I believed success conferred worthiness, that I was prestigious and making a lot of money. I was singularly obsessed with those things." And when Goldstein had her first child at thirty-one, she still adhered to her faith in this model of success: it helped that she occupied a "really high-profile" media job. "I thought it was only self-doubt that would hold me back," she observed.

Within six months, though, everything changed when Goldstein "brutally" lost her job. Her "identity and self-worth were in a tailspin: if I wasn't a successful professional, who was I? I must be the worst new mom worker in the world, a person who loses their job just after maternity leave, I must be the worst failure."

She was embarrassed when her own self-sufficient-working-mother story started to fracture. "I didn't immediately question the ethos," she said, "I took losing my job as an example of my own personal failure." As Goldstein frames it now, "The personal responsibility narrative cut both ways: if you alone are responsible for your success, you are also responsible for your failure."

At some point, she began to question her assumptions and values. And somewhere between losing her job and the shifts in the political firmament, including Trump's election, Goldstein started to feel her beliefs altering. She found herself doubting the version of success she had previously embraced. She also wondered about the female entrepreneurs she had followed and whether they examined how their own inborn or early privilege sedimented their advantage. She felt she had, in the past, "ignored the social forces that allowed me to get where I was."

She was also newly aware of how the model of C-suite-motherhood didn't linger on how caring responsibilities that come with parenting—and sometimes at the same time looking after an elderly parent—didn't fit in with the pursuit of this kind of career.

This was when Goldstein finally threw out the I'll-do-it-myself-working-mother idea and started to talk to lots of other mothers about their condition. It became hard for her to even mention *Lean In* "without grinning ironically."

Goldstein now lives in Durham, North Carolina, with her family. She hosted a podcast (my organization supported one episode) that was devoted to mothers' struggles in the labor market (seeking to challenge "the status quo of motherhood in America"), up and down the income gradient. She has become radical about her former ideals, especially in the light of her parenthood, rebuking the self-made worldview that once defined her. "Even noting something like 'Look how much venture capital she raised' was a way of judging a woman's success by male-dominated games," she said. She also acknowledged that by the same token, "a huge amount of my early success was related to economic privilege and to white privilege."

I recognized my own experience in Goldstein's account, although without the *Lean In* accoutrements. I, too, had forced myself to try to climb high when I was a young woman. I hadn't bought into the alternative be-more-chill ideal that was also on offer, partially because I felt I always had to strive to survive, for my bank account and my very existence. Eventually, though, I recognized not just the extent of the problem of this kind of self-punishing work ethic but also both the social class insecurity as well as, paradoxically, the social class privilege that fueled these ambitions. While the contest of American achievement was in fact tilted, to some extent, to benefit people like me—I knew what to aspire to and what behaviors were permissible while striving—I also didn't feel protected enough financially or professionally to ever stop hustling. And that didn't change when I had my daughter. In fact, the need to support our family in a major American city made my constant striving and endeavoring all the more effortful. Now the myth that I had to accomplish the best I could on my own had serious urgency.

In the early months of the pandemic, in 2020, an isolated drudgery engulfed me, as it absorbed so many others. The refrain that women should "be a boss" at work echoed oddly during that time, like its message was an old greeting card, a cliché that had become nearly illegible. Most parents during COVID had to be a boss in a way that abraded them—to be "the boss" of your kid's remote learning, say.

I remember early in the pandemic preparing dried lentils from scratch while editing professional writers and then running over to play in-house tutor for my daughter.

It wasn't a coincidence that during that time I became even more enamored with the philosopher Kathi Weeks, author of *The Problem with Work*, who tries to reimagine work to include a "postwork" possibility. While the individualism story tells us that hard work leads to limitless possibility, during the pandemic we learned what we already knew—that often the opposite was *actually* our lot: being stuck; laboring and receiving not enough in return; depending on others and endlessly being depended on. Perhaps exploring postwork, what Weeks suggests, could offer us a solution—at least for a while.

=====

In 2020, I dreamed of a parents' mass movement replete with maternal lobbyists, organizing and fighting for support. It reminded me, in both the syrupy slowness of time and the utter absence of others to assist us, of my daughter's early infancy.

I was not on my own. Other parents informally began collective efforts, including in the much-maligned educational "pods" families set up and independent learning groups they created as well. The growing dissatisfaction was also more publicly codified when a group of fifty female leaders ran a full-page ad in the *New York Times* calling for the government to implement a Marshall Plan for Moms. It was started by Reshma Saujani, the founder of the organization Girls Who Code. The group urged the government to pay mothers a $2,400 monthly stipend for caretaking labor and for affordable childcare as long as the pandemic lasted.

At the same time, Monica Scott and other parents I talked to were involved in less glitzy but no less revolutionary efforts, small tranches of parents engaging in a sometimes embryonic movement, which included older groups like ParentsTogether and advocacy events like the punning Strolling Thunder. One of the latter's events connected "babies and families from all 50 states and DC to their Members of Congress." One of the biggest parent-centered groups, MomsRising (with a nearly $9 million annual budget in 2017 and a million members), had "fought hard to get real [political] acknowledgment that childcare is not an industry but an infrastructure, and without it parents cannot get to work," as its CEO, Kristin Rowe-Finkbeiner, told me. In 2020, MomsRising had directed 406,632 citizens' political responses to Congress from member mothers from the time the pandemic started. Its members made 41,790 phone calls to representatives supporting COVID-19 relief packages in the first half of 2020. I was reminded of how around fifty years ago, Swedish feminists lobbied zealously for the expansion of childcare to great effect.

In the end, it was all of these parents' groups, plus the Marshall Plan for Moms folks, that in March 2021 helped get the American Rescue Plan, with its massive $1.9 trillion in spending to pass in the House of Representatives. It was hailed as the biggest policy gain for children in the United States in twenty-five years, then passed by both chambers and signed into law. We seemed to actually be building what domestic work avatar Ai-jen Poo called "care infrastructure." The American Rescue Plan had childcare, education, and tax-credit provisions. Parents would get tax credits for childcare work they did in their own home, up to $3,600 per child under the age of six. There was an additional $15 billion available to help low-income families access childcare. In addition, the stimulus package offered an unemployment insurance supplement of $300 a week through the first week of September. By October 2021, parent activists were marching in DC to illustrate how critical childcare, paid family leave, and child tax credits were to them.

Furthermore, the Build Back Better infrastructure package had set aside $400 billion in support of childcare, early education, and wage raises for childcare workers. It expanded childcare and pre-K and ensured copays for day care for the children of middle-class and working-class parents, paid for by the government.

It seemed as if Mother Goose had visited the White House and whispered in the ears of those in high command. If only she—and they—could make these new cushioning conditions permanent. The plan had been spearheaded by two Democratic female legislators, Congresswoman Rosa DeLauro of Connecticut and Senator Patty Murray of Washington, both in their seventies, who were familiar with the lot of parents and had been pushing for a plan like this for more than twenty years.

The plan also extended tax credits to cover COVID leave, so families could care for loved ones who were ill if their care was interrupted.

The stimulus bill was part of "a consciousness raising" that came out of the pandemic about what workers like these women experience, said Kathleen Gerson, a professor of sociology at NYU. "The most cataclysmic events of the twentieth century, like the Depression," said Gerson, led to similar "silver linings," like the formation of Social Security. From the vantage of 2022, after many of these hopes have been stymied, it may be hard to remember that 2021 had also seemed a similarly momentous time—post-Trump, and then (we thought) edging toward postpandemic, when real talk about raising the minimum wage and government-funded childcare could happen and when there might even be a reset. Perhaps there was an opening for rethinking the role of government in the family, and perhaps for new solutions (which were often old solutions) involving more state aid. There was also a welcome shift within political discourse to more family-aware language. The Biden administration, as I noted earlier, began to use the phrase *human infrastructure*, interweaving common sense and structural aspiration.

When I spoke to a handful of mothers about what they were spending these new stimulus bill dollars on, their responses, and the RAPID-EC Project, the survey I mentioned earlier, showed us that families with one child under age six were by and large spending their stimulus checks to pay off debts. This was especially true for Black and Latinx families. Those getting stimulus checks and unemployment also spent them on food, utilities, and home supplies. There was an emotional component to the assistance as well. (When our family got our childcare tax credit checks, there was also the warm flush and a novel sense of confidence that our country cared about us and about our family; that America was working for families and knew what we had been through.)

Scholars remained hopeful that we would emerge from the pandemic with a renewed commitment to a caregiving infrastructure; that the stimulus bill and related governmental payments could ultimately cut child poverty in half.

"The most important part of the stimulus was the direct payments—the check," Teresa Ghilarducci, an economics professor at the New School, wrote to me. Unemployment insurance benefits came after, allowing mothers not to look for work instantly, when their kids still needed childcare. Higher income workers were more likely to keep their jobs and were also more likely to supplement their children's education with their own resources; for lower income women, the child tax credits were particularly important and not just economically (one older study from 2011, by economist Raj Chetty and other scholars, found that tax credits raise children's test scores, creating better longer-run outcomes).

Another crucial recognition that was baked into Build Back Better, at least as it was initially proposed, was seeing the care economy itself as fundamental. Nancy Folbre, an economist who studies the care sector, had argued for twenty-five years that it is a key part of the economy. In May 2021, she told the *New York Times*, "I often say to myself I'm glad I lived this long so I can say maybe I had a point."

A few months later, of course, some of these same would-be advances would be blocked by a partisan Senate. Congress was negotiating, and conservative Democratic senators Joe Manchin and Kyrsten Sinema were holdouts.

The battle continued to make these and other federal payments of their kind not simply gauze wrapped around a catastrophe but part of our enduring social fabric. But the arguments against, say, tax credit payments were not just coming from pinchpenny senators. They were also coming from actual parents who had drunk deeply from the individualistic chalice, among them a focus group convened by the right-leaning Institute for Family Studies, which considered these payments, which were given to roughly 90 percent of Americans, "unfair": "It's going to just allow them to abuse it, not have to work."

Even confronted with these steps back, there had clearly been steps forward as well, among them, a mass recognition that family-support measures can help siloed parents survive and the self-made myth as it pertains to families can be dangerous, particularly in a crisis. Finally, the ways in which parents—mothers in particular—are allocated money by our government can be reimagined. They were created by human beings, not gods, in the first place.

Monica Scott wanted less evanescent help. Scott joined Fight for $15, which campaigns for better wages for lower-income workers. "I am so excited: I love it," she said of her new avocation. She is trying to get bigger weekly paychecks and more protective gear for her job at Amazon. "I want to fight against unfair wages and also for child-care," she said. She was having a fourth child and there was not likely to be an easy track ahead for her. Nevertheless, she can imagine, she said, having her life be better, or as she put it, "better up," an idiom all her own.

10

The Con of the Side Hustle

VANESSA BAIN SPENT much of her day shopping at supermarkets. She wasn't haunting their garishly lit aisles to purchase food for her husband and daughter or the three elderly relatives she was regularly in contact with. The then-thirty-six-year-old was doing it to support her family in costly Silicon Valley.

Bain said she liked being a personal shopper before the pandemic. She saw it as offering adequate money and flexible hours, and it fit well with her style of work. Bain was, by early 2020, seeing fewer of the pluses of this vocation The benefits had dwindled, and at work she had a far-increased exposure to a dire virus, as did her husband, who was also a professional shopper. She also started to feel she wasn't being treated fairly.

By March 2020, she was disinfecting her phone several times a day, partly because she had to hand it to customers to sign for orders. Her personal reserves—that she had paid for—of isopropyl alcohol, hand sanitizer, and antibacterial soaps rapidly depleted. The civic emergency was taking its toll. Bain had worked for various companies, including Instacart, Uber Eats, and DoorDash. Armed with baskets and carts, she searched for black beans or frozen macaroni and cheese, walking the same aisles more than once, purchasing goods for people at home, at a time when little was known about whether boxes were contaminated on contact or whether masks would protect against the air in crowded supermarkets. Afterward, Bain drove the bags of foods to strangers' homes and often interacted with them on their doorstep or hallways, another risk vector. Jobs like Bain's at these app-based companies often didn't come with health insurance, despite the fact that Instacart, as of October 2020, was valued at $17.7 billion, and the company's entire business depends on its shoppers. This number had gone up to $39 billion by March 2021. Instacart's bootstrapping owner, Apoorva Mehta, was often admiringly cited in the press as an "immigrant founder" who provides "essential services," giving "jobs and a lifeline to many," but the astronomical valuation of his company contrasted sharply with how pinched the contractors' personal budgets were. Mehta, "the serial entrepreneur" behind Instacart, was, by June 2020, according to *Forbes*, "leading a $13.7 billion unicorn after demand for grocery delivery skyrocketed during the coronavirus pandemic," writing approvingly that Mehta's grocery delivery service was tamping down the efforts of "rebellious workers."

The CFOs and CEOs had their individualistic mantras, yet in truth these potentates depended on masses of people like her.

Bain, when I first spoke to her, was working a gig job that is sometimes referred to as a "hustle," or a "side hustle," if it takes up a smaller slice of a person's overall hours. While hustling, like "pulling yourself up by your bootstraps," had become an aspiration, sold as one element of how one achieves the American Dream, it was also filled with difficulties.

There were an estimated fifty-seven million gig workers, shopping, fixing, driving, and delivering. Before she felt put at risk, Bain said she loved what she did, that when "you shop for the same customers, so you establish relationships." She worked at her own pace and was able to pick up her daughter from school and felt like the job suited her. She also felt that she and others in her line of work were purveying basic social services for some customers, including those who were disabled or housebound. Shopping seemed to her to be something of a calling, a valuable service. Bain also felt she needed to do extra for her companies' disabled customers: "We are not going to punish a customer who cannot lift their stuff." Bain had heard "countless anecdotes of people saying, 'I rely on this gig shopping service as I am housebound' or 'I have an immunocompromised child that I cannot take to the grocery.'"

She and her husband didn't make much—she and her family were living in a unit in the back of her grandmother's house with an adjusted household income of $28,000, making her family among the poorest in her area. Nevertheless, while the 2020 rationalist appellation of the "essential worker" became a social and legal category, one that was sometimes deployed cynically by politicians for their own gain, Bain *really* felt essential all along for her work, getting a sense of dignity from her labor that came from the inside.

At the same time, by the spring of 2020 she felt that she was not being treated fairly. In addition to not having medical coverage, she did not have unemployment insurance, workers' comp, minimum wage, or safety protections. These were particularly necessary, as Bain often carried heavy bags and boxes to and into people's homes, up driveways and stairs, in all sorts of weather. Sometimes, she said, she was the only person her clients saw in a day. Bain had been admitted to the emergency room before, just from doing her job and shopping for companies like Caviar and Uber Eats. Before the time of COVID, she wounded her forehead after falling down while delivering groceries for a client. She said she is not the only one. "Look, a shopper also cracked her skull open and broke her leg when she slipped on ice helping an elderly woman with her groceries and was in a coma for three weeks," she recalled. She knew this because she was one of a handful of workers organizing daily on behalf of the army of Instacart shoppers who labor without a union or other job protections. She also understood it intimately because of her own physical compromises. She had chronic sciatica, an agonizing nerve ailment. "Instacart is their lifeline, yet many shoppers themselves have disabilities and do this kind of work through chronic pain," she said.

What the pandemic had done, however, was further clarify the risks of jobs that are part of what we might call the hustle economy. The "main hustle" and its sibling, the "side hustle," may well function as concepts that disguise unstable working hours and a lack of bargaining power as liberation. While Bain and others found fulfillment in their full-time shopper or driving gigs, and their part-time versions, these lines of work can often be prime examples of inequality within labor.

On Facebook and Reddit, the company's workers posted about being denied pay for small failings—not being able to get a COVID test at one point, for example. A post on Reddit, "Has anyone received their covid pay from Instacart?" in the summer of 2020 led to a cascade of depressing responses. One worker said they had had "covid for weeks" and had filed for "covid pay." They had been told they would be paid "within 14 days" but were "currently on day 14 with no pay/contact for weeks." The furious worker continued: "Pretty crappy for a 14-billion-dollar company to not even pay their workers a fair sick pay in a pandemic they're profiting off of."

(There were many such episodes during the pandemic at companies like Instacart and those far bigger: the New York attorney general's office eventually filed suit in 2021 against Amazon: the complaint noted that there were eighty instances of confirmed COVID infections, but the company failed to close any part of the facility in response.)

Bain was mad about it, and she knew she would have to do something about it. She posted heckles like "We really doing this dance again, Instacart?"

She remained vigilant. A few years before, Bain, a college graduate, had started an Instacart grassroots advocacy network, first through Facebook and the nonprofit group Gig Workers Collective. How could companies keep them as "independent contractors rather than employees with full benefits" when a disease was rampant, she wondered out loud.

═══

One of the ways that gig companies have been able for years to define their workers as independent contractors is by exploiting the allure of individualism. The contractor has their freedom, the thinking goes, so why would they exchange that for a full-time job with benefits? But another culprit emerges from popular culture: the trendy pidgin to describe contract work tends to glamorize such work. Our jobs, for example, are now "flexible," because we are the ones contorting ourselves to work at all hours, or we were professionally "nimble" because we were trying to survive on freelance gigs. The lingo around living paycheck to paycheck routinely tried to make the dreary carousel of contemporary life sound more fun. "These words have gained a strange kind of prestige from downwardly mobile, college-educated tech workers," said John Patrick Leary, the author of *Keywords: The New Language of Capitalism*.

It was part of what Reddit's founder, Alexis Ohanian, has called "hustle porn," cheerleading multiple jobs and their backbreaking exertions, giving a cheerful spin to the full 30 percent of Americans who do something else for pay in addition to their full-time jobs, according to an NPR/Marist survey. Hustling and nimbleness and the like implied, however, that there is a slinky joy in picking up gear at a studio and dropping it off across town or selling CBD oil part time.

Once called drably "another job," these more chic words gave instability—that fewer than half of American adults, just 47 percent, say that they have enough emergency funds to cover three months of expenses, according to a survey conducted in 2020 by the Pew Research Center—a shiny gloss.

The con of the side hustle happened when an NFL coach admiringly remarked that doing your job right means waking up at 3 a.m. with "a knot in your stomach, a rash on your skin, are losing sleep, and losing touch with your wife and kids." It was in the exhortations of former SoulCycle CEO Melanie Whelan, who pushed "hustle culture" to one and all. "Hustle opens the doors of opportunity," Whelan once said, encouraging her followers to work long days with the exhortation to "rise and grind." The urban legends of side hustling at its worst included one Lyft driver who continued to pick up riders after she went into labor, then Lyft-ed herself to the hospital to give birth, and T-shirts with the slogan "9 TO 5 IS FOR THE WEAK." It was in a set of risible "hustle culture" memes extolling the "grindset." One maxim: "You can't make excuses and money. Which is it going to be?"

Commercial websites like Side Hustle Nation also were part of it: as Nick Loper, the site's chief side hustler, exhorted, "My escape route was a side hustle business I built in my spare time—and you can do it too." The site Medium had a whole Side Hustle Club, with the motto "Where the outcome is income."

This "grind" vocabulary for work can even be heard in a new rendition of that famous song by working-class patron saint Dolly Parton, where she twanged in praise of working on the side and in the off hours: "Workin' five to nine . . . a whole new way to make a livin' / Gonna change your life," all in the service of an ad for Squarespace, a Web-hosting platform, "Be your own boss and climb your own ladder."

The idea haunting this jargon is that if you are coordinated enough, clever and cool enough, you can bootstrap your way out of two jobs and, ultimately, into achievement.

In truth, the side-hustle life wasn't all it was cracked up to be, and not just for the typical gig workers but for part-time workers in general.

For instance, at the onset of the pandemic, Brenda Madison, in Laguna Beach, California, lost not one or two but *three* different gig jobs—from doing graphic design to working at Athleta, and she put it like this: "We are all running in place as fast as we can to stay the same, to quote *Alice in Wonderland's* Red Queen. Never did I think that a medical injury or unexpected repair could bankrupt us." Nicole Braun, a fifty-five-year-old adjunct professor of sociology in Chicago with one grown son, also shared with me how for decades she had taught classes at multiple colleges to earn a living—as many as ten classes a semester (teaching in summers, too), but at the beginning of the pandemic, work dried up: student enrollment was down at the colleges where she primarily taught. She sent out "literally hundreds" of job applications but to no avail. She grew frantic and submitted her application in May 2021 for unemployment insurance. Then she waited for the checks to arrive—and waited some more. Like other contingent workers I talked to, Braun's economic difficulties didn't always meet the naked eye. It was as if she and others were, proverbially, hidden under the paper flaps of an Advent calendar, where each box covered the faces of who gets hurt, and how, in this country.

After dozens of letters and calls, Braun also got much of the money she was due, though by her estimate she is still owed $1,800. As of my last exchange with her, she was dog walking and teaching classes at less than $3,000 per class, but getting ready for another season of contingent drought, "begging for work from the places I teach at, but there is nothing so far since enrollment is down.

"We live in a culture that values individualism, and thus we are socialized to see through that lens; it causes much shame," said Braun, who is a sociologist after all, of her unreliable work schedule. "There is a part of me that still feels I did something wrong."

In addition to side hustling's secret and melancholic instability, scholar Tressie McMillan Cottom observes, hustling can serve as a "kind of racial theater." American culture "applauds the hustler" for "striking out on her own," even though she may be doing so because she is excluded from traditional and secure employment due to bias. Digital platforms and companies "celebrate grit and urge us to 'respect the hustle,'" she writes, but this glorification covered over the fact that it's usually the already privileged whose side hustles add up to more than the sum of their parts.

Hustling's implied glitz and energy in fact derived from a hip-hop genealogy, Leary said. It had been alchemized in rap from the 1990s and 2000s (the song "I'm a Hustla") from something that Black people in particular had long needed to do as they worked a job—or multiple jobs—for which they were not properly compensated and the only way to make ends meet was to work an additional job.

———

One of the companies that is most guilty of mystifying the side hustle has been Uber.

It even sold itself as a hustle opportunity in advertisements as well as to its drivers through algorithms that are always trying to keep the workers working harder while at the same time nurturing a feeling of hope. Legal scholar Veena Dubal, in one of her accounts of the company's corporate mind control, writes of how app-related communiqués encourage drivers to work with more enthusiasm by using the "expertise of behavioral psychologists." Drivers "receive alerts about 'surges' and 'personal power zones,' which they understand as increased demand in a particular area," she writes. Words like "promotions" and "badges," and "surges" and "personal power zones," are, like "side hustling," meant to shore up pride, "designed to create an enchanting sense of possibility."

Saori Okawa, who logged 3,700 rides in a single year driving for Uber, felt this hope working on her, at least at first. Uber was the only job she could get that would work with her school schedule.

At first, Okawa would leave her apartment and her three roommates in San Francisco's Japantown at seven in the morning to get the best of the early morning commuters. She experienced freedom, said the then-forty-one-year-old. "It's why I am here and not in Japan: they are very narrow-minded. People think differently here—difference is OK." In Japan, women are not supposed to drive for Uber, she told me, because there is "strong sentiment it's a masculine job: my parents will have a heart attack if they know I drive for them." Okawa felt there was an element of societal and gender rebellion in her vocation. It fit in, she thought, with her overall nonconformist aura, her flouting of the rules of traditional femininity: Okawa's hair was always clasped in a ponytail, for instance, and she was always clad in black athletic wear. As she said, "I wear mostly sporty: I am not making it fashionable." She is an ardent Catholic, having converted from Buddhism four years before after having been married to an American. She had, though, jettisoned what she found to be a stifling marriage, which felt far too brassbound.

She took classes in social work and did outreach through her church, feeding San Francisco's unsheltered and visiting—with her psychology classes—the tent cities that dotted downtown. She felt that the people she met there were "broken" in the way she was, with what she describes as a fractured past and present. "When you are in brokenness, you don't think there's anything you can do," she said. "I know that feeling."

Okawa also drove constantly to areas where the unhoused slept in makeshift structures and sleeping bags, places in the city and its environs she probably would never have seen normally. Divorced and single, she rarely saw her friends because she was always working or studying or attending church services or, as she captioned her experience, "No life." And while on a good day she'd make $180, she also had to pay to rent the car, which was $200 for the week, and then also gas, which averaged $90 per week. She was cashing out every day so she could put gas in the car every night. "I had to work constantly," she recalled. After a year, Okawa started talking to other gig workers about the way these app-based companies treated their contractors.

Uber and Lyft still classified ride-hailing drivers as contractors, so these companies didn't pay into state unemployment funds. A University of California at Berkeley study found that they would owe the state $413 million, for the period from 2014 to 2019 alone, if they had correctly classified their drivers as employees. And when these workers are able to access funds, other taxpayers and the state foot the bill, and the drivers themselves are the ones paying taxes.

For Okawa, the fear around COVID-19 had only intensified her frustration with the labor conditions of her driving. "A person should be treated with respect," Okawa told me. "During the pandemic I realized how much we needed the sick leave." The companies didn't provide any of the protections from COVID they promised. At best, ten single-use masks arrived by mail from Uber, after a month. (Remember, Uber bought Postmates for $2.65 billion *during* the pandemic.) Her coworker got COVID. As Okawa put it, "Why can't we have an organization that protects us, why can't there be a union, why do we just have to say yes?" Postmates "gave me just one piece of cloth and Instacart gave me a very small hand sanitizer and one mask," she recalled.

Meanwhile, the Uber "Greenlight Hubs," or worker headquarters, that were supposed to carry hand antiseptics, all closed at the start of the pandemic, and some never reopened. With "absolutely no masks," Okawa said, so the drivers had to share what face coverings they had with one another. The customers were nicer, though, at least at the start of the pandemic. They'd tip better and were more thankful when she showed up.

Soon enough, Okawa's always-difficult business flatlined. Sometimes she would make as little as two dollars a ride, sometimes even during a rainstorm. Okawa, an enterprising woman, then rushed to join other apps, like Postmates and the companies Bain worked for, DoorDash and Instacart.

She started to prefer the grocery delivery because it seemed safer than ride-sharing during a pandemic, but her days were still long and she'd leave early in the morning to pick up orders from Walmart, because the drivers had started to notice that the pay for these same jobs went down as the day went on. She was still coming home every day twelve hours later. "I don't know if I am making minimum wage," she said.

When I spoke to Okawa and other gig workers about their work experiences during the pandemic, their treatment by their companies reminded me of what philosopher Judith Butler had written about "grievability." She used that word to describe those who are deemed valuable enough to be mourned. It also described those people who don't tend to be grieved for and don't make the grievability cut. The hierarchy of who is and who isn't lamented and who is and who isn't protected from risk get at the deepest part of inequality.

=====

This cute but misleading phrasebook of self-reliance now cloaks our stressful work lives. How do we challenge this framing?

To start with, we can take a hard look at the term "side hustle" itself and ask what it really does to our minds and bodies. We can decide that it is an insidious term we won't use again, part of a falsifying new parlance that we should overthrow in search of a more truthful language. This is a much-needed renovation of language—to interrogate the lingo of a society and "change as we find it necessary to change it, as we go on making our own language and history," as Raymond Williams, the late British cultural theorist, puts it.

We could also start emphasizing other words and phrases, among them the word *security* when it comes to work. Seeking out security instead of side hustling would mean prizing safe and comfortable workplaces, long career tracks, and safety nets.

We might also reclaim one word of the pandemic, *essential*. Leary, the author of *Keywords*, a contemporary book that riffs more humorously on Williams's original book that had the same title, *Keywords*, but with the added subtitle *A Vocabulary of Culture and Society*, wrote to me that the Department of Homeland Security's list of essential workers includes landscapers, exterminators, credit card factory workers, and armored-truck drivers. That's not to say all of these jobs are *inessential*, but they are certainly a watered-down version of the word's original meaning. If we were to reappropriate the word *essential*, we would be better positioned to reward those who are doing work that was and still is essential to our survival.

The other way we can fight the con of the side hustle is by organizing, which Bain herself did. She thought: *If overtime work is paid at a much higher rate, shouldn't life-risking work during the virus time be much higher paid as well?* So she began to canvas for hazard pay for the Instacart shoppers. Workers who performed a key service during the pandemic wanted a small raise for the risks they were incurring. The companies saw the PR upside of it and publicly announced they would be offering hazard pay. A month or two later, though, they clawed that offer back when no one was looking.

Eventually, Instacart shoppers staged a walkout in October 2021 over not seeing their pay improve. (Instacart's director of shopper engagement and communications blandly commented to the publication *The Markup* about it, "We take shopper feedback very seriously and remain dedicated to listening and learning from our community to improve the Instacart shopper experience.")

In the face of this agitating, Instacart offered free telemedicine visits for workers who had been exposed. But once again, these forward steps were often followed by backtracking. Mehta held firm: he was praised, in fact, by *Forbes*, for not rolling over to his contractors. Instacart terminated two thousand of its roughly ten thousand in-store shoppers in March 2021 to save money, leaving them with no other source of support.

Okawa was radicalized by feeling abandoned by responses like the above from the companies she worked for. That emotion led her into more collective forays.

"I am really glad I am a part of this movement," Okawa said: her activism was saying no to the hustling mentality on a fundamental level, even if she still had to continue in accord with its practical demands to survive.

Okawa, for instance, became involved in the opposition to California Proposition 22, the Uber- and Lyft-funded ballot measure that worked against California's AB5. AB5 had been a huge victory when it was signed into law in 2019: it gave employee status to gig workers and made it much harder for the megacorporations to claim their workers were independent contractors rather than employees. Okawa also joined a local organizing group, We Drive Progress. The group defined itself as a collective of "app-based drivers," those "behind the wheel of every trip for Uber, Lyft, Instacart, and beyond." Their protests were aimed at "fair wages, benefits, and our unions" as well as earning a "fair share" of "the billions these companies and their investors pocket." She took part in rallies in front of Uber headquarters with other activists. These organizing efforts seemed to have the intensity of secular rites, and the efforts in themselves can offer strength. It kept her and others from becoming just one of the workers who were like underpaid data points, what the theorist Franco Bifo Berardi aptly terms "small parcels of nervous energy picked up by the recombining machine."

Okawa and her colleagues also took note when other companies with many contractors like her coalesced in mass organizing, as when Amazon workers went on strike in New York, Michigan, Illinois, and Washington and famously organized in Bessemer, Alabama (the union was ultimately voted down in April 2021 but had surprisingly strong support). Okawa also saw that the activism wasn't limited to traditional union efforts, but included coalitions like Amazonians United and a mesh of activist, pro-labor, and social-justice groups.

Taken together, these sorts of worker campaigns subverted the mindset that we individualize every second of our existences.

Okawa also now sometimes had coffee with her fellow drivers

and then shoppers in between rides and deliveries and tried to recruit as many other delivery drivers to gig work activism as she could. In 2021, her efforts helped get a city ordinance passed to pay for the cost of sanitizing for even part-time contractor gig workers. She found this result exhilarating. Collective campaigns could sometimes take ordinary, lonely geographies and transform them into sites of possibility: this seemed true for Okawa.

She, along with those like her, had joined—and leaned on—one another in a fashion that was both idealistic and sensible. They were acting on something they had long understood, despite constant messaging to the opposite, that no one can hustle alone for long.

IV

TOWARD A NEW AMERICAN DREAM

11

Class Traitors

Between the worker and the millionaire
Number provides all distances.
—DELMORE SCHWARTZ

IN 1986, CHUCK Collins, an heir to the Oscar Mayer meat-manufacturing fortune, gave away his fortune. He was twenty-six at the time, and the amount was roughly half a million dollars—in today's dollars 1.2 million, although he estimated that if he had kept and invested it, the money would have appreciated to $7.75 million by 2015.

It was what he saw in his first job out of college that led to this startling act of generosity. He was organizing mobile-home owners who were trying to buy the property where their homes were situated, attempting to purchase the land from private owners in western Massachusetts. Their ultimate goal was to turn a whole park into something they owned cooperatively. Collins sat in a double-wide trailer with its owners and other tenants as they met for a moment of great anxiety. He knew that most residents' finances were pinched at best and one had no savings at all. Collins and the residents spent a day anxiously trying to cobble the funds together. Seeing these people struggle, in a state of emotional disarray over finances, had a transformative effect on the young man. As he put it, it was a front-row seat to economic deprivation.

Finally, by evening, they had come up with the necessary $35,000. Collins was never the same.

During the months when he helped the mobile-home owners, Collins told me, he was aware that he himself was not suffering as they were. In fact, he had been compounding wealth of his own, without lifting a finger. Recognizing this discrepancy, he felt self-disgust and decided he no longer "wanted to benefit from a system of inherited wealth where some start with so much and others have none."

He told his father his intention to give away his inheritance and informed a wider circle, too. He was told by others that he was "going to regret this" and that his "brain hasn't developed yet," making his choice "irreversible." Some of them thought he was a lunatic for turning away from his family's immense riches. As his father used to say, "Bringing home the bacon has a different meaning in his family," as Oscar Mayer is quite a name in lunch meats.

Collins felt he knew what he was doing. He no longer wanted to possess a fortune and for it to influence his decision making.

Collins was sixty-two when we last spoke. In the intervening years, he had gone on to become an expert in wealth inequality, and that was how I had met him: the site he was connected to, Inequality. org, had collaborated with my nonprofit.

He lived a seemingly comfortable, middle-class life—retirement savings, home in Boston, summers at a farmhouse in Vermont. Collins's two grown children also refused their inheritance from the family fortune, choosing to pay for college with loans that they're still paying off.

You could call him an antibootstrapping militant. Or a consciousness-raising agent for the wealthy.

Collins is modest in appearance, a White guy with a middle part who prefers comfortable cotton shirts to suits and ties. Yet for all of this gentle nerd-ocratic-ness, his calm, reasoned sentences can't cloak his doctrinal intensity or the certitude of the group he represents, the Patriotic Millionaires. It's a nonprofit consortium of the self-aware rich or, as I call them, the "transparent rich." Their website describes members as "traitors to their class" and elaborates: "Patriotic Million-aires are high-net-worth Americans, business leaders, and investors who are united in their concern about the destabilizing concentration of wealth and power in America."

Pivotal to their work is the idea that a primary barrier to those with privilege becoming more societally generous was that they were not publicly honest—even with themselves—about their money and where it flowed from. They were exposing those who pretended their wealth was accrued as a part of a fundamental order rather than at least partially the outcome of the top of society crushing the bottom. As the writer Anand Giridharadas put it in one interview, it was hard to find extremely wealthy people "willing to do the only thing that is actually going to get us there, which is fighting for the kind of systemic change that would reduce their own power." The Patriotic Millionaires and those like them wanted to be the exception, in part by emphasizing openness about their finances.

They were working in opposition to the usual rules of their caste, whose rituals were often secretive, who had their own lingo, eti-quette, schools, neighborhoods, and even seaside towns that they visit seasonally. The Patriotic Millionaires were modeling what might be possible if America's overclass punctured the idea that they were al-ways deserving of their outsized fortunes. And they are starting to gain traction.

Part of the Patriotic Millionaire process—along with that of allies of the group, which include organizations like Solidaire and Resource Generation, is to "come out" about the inherent privileges that al-lowed them to secure their riches and encourage other wealthy people to do the same. By doing so, they are exposing the fact that the same names recur on the ballot boxes at each election and in the ranks of the Fortune 500 companies announced each year. In addition, they are making it their mission to change the scarcity mentality of the rich toward what one scholar calls a "generosity mentality." They are also casting themselves as part of a network rather than sui generis kingpins, and finally they are publicly recognizing that their group overall is a tiny minority. Ultimately, by doing all this, they are attempting to bridge this income and asset inequality chasm.

The transparently rich activist members of the Patriotic Millionaires, and groups like them, believe that if more wealthy Americans "came out," great things could happen. Rachel Sherman, the author of a book on the ultrawealthy, *Uneasy Street*, told me that she believes that the Patriotic Millionaires' efforts to "reframe self-interest" by questioning the rich-deserve-theirs story could make a difference. It isn't just talk. It's not what Ayn Rand might have called contemptuously "an orgy of self-sacrificing" either, although it's getting closer to that state.

I got a taste of what this "coming out as rich" spiel was really like the first time I spoke on the phone with Iimay Ho, the (now former) executive director of Resource Generation, a nonprofit working to organize wealthy young people with class privilege to work to relieve inequality. She immediately came out to me about her financial status. Ho has personal access to assets of $100,000 plus her $370,000 apartment, and then made $68,000 a year. In New York City, these figures don't necessarily read as rich, but they do, demographically, put Ho in the top 10 percent of those in her eighteen to thirty-five age bracket, she said, or even the top 5 percent, depending on how you slice it. The cost of her apartment might not sound astronomical to some ears, but Ho mentioning it is part of how new resource-aware people talk about real estate in particular as part of the long litany of racial economic injustice: "Homeownership is a marker and a part of a huge transfer of wealth." Resource Generation maintains that wealth is defined differently by different people, but they note that US households at the ninety-fifth percentile have a combined income of $148,000 and assets above $1.3 million; many of the nonprofit's wealthy young recruits have considerably more.

My disarming chat with Ho was meant to peel back what often cloaks the wealthy, especially those who have inherited their fortunes. I've long noticed among my creative-class social set a mysterious silence around family money and being well off in general. Ho's transparency was refreshing in comparison. Ho and Collins and their coterie find this transparency to be urgent and necessary. Our individualistic mindset has political side effects, including the way silence about privilege and poverty separates people from one another and prevents them from uniting for change, or even voting along the same lines for maximum power. They shouldn't crow about how hard they work and how much they deserve their opulence. In return, this kind of self-aware activism offered a cleansing, as if allowing the members of this self-revealing-rich group to burn sage all over their own affluence.

There were reasons why they don't want to come out as such in the first place. They could be protecting their assets, believing they will be asked to donate to causes or individuals at every turn, or they might fear being taken advantage of. But privileged people's aversion to talking about it was just one way that the roughly two-hundred-strong Patriotic Millionaires have their work cut out for them, along with their tagline "disrupting the narrative of deservedness."

They had hundreds of years of ideology to combat in their quest. Changing the siloed mentality of America's wealthy involves seeing their success as dependent on the survival of those who are less advantaged than they are. "You need good public health if you are going to accumulate under capitalism," Sherman said. "You have to have workers who are alive."

The class-traitor rich have a very systemic change in mind for their wealthy brethren: paying their *fair share* of taxes, including the wealth tax. Giving to causes rather than paying taxes shouldn't be an option. Taxes, as opposed to gifts from the rich, allow elected officials to decide where this money should be spent, and not just on pet causes that can offer tax benefits for the rich, dubious loopholes like carried interest. (The next time you meet a philanthropist with, say, a private skating rink and a home with pergolas, ask them about "carried interest.")

In this way, the Patriotic Millionaires closely follow the blueprint of the University of California at Berkeley economists Emmanuel Saez and Gabriel Zucman, whose study of wealth taxes influenced Senator Elizabeth Warren's own proposal. That wealth tax would place a 2 percent annual levy on wealth between $50 million and $1 billion, and a 3 percent tax on wealth over $1 billion. Biden was also actively pressing for greater taxation of Americans who made over $400,000 a year. The idea behind that and the Ultra-Millionaire Tax Act, which would tax the 0.05 percent wealthiest their ostensible "fair share," is that these increases would overall generate $3 trillion over the next decade (which in itself might seem a relatively modest amount): a corporate tax on companies' profits above $100 million would bring an additional $1 trillion over the same ten-year period. (In 2022, Biden, too, called for a 28 percent corporate tax rate.)

Tax activists, the transparent rich among them, were also attempting to raise the corporate tax rate to 28 percent from 21 percent, as well as taxing long-term capital gains for the richest households. As tax reform advocate Erica Payne told me when the pandemic was raging, that period could become a prime "opportunity to un-rig the economy and wrestle the control away from the self-interested billionaires." (As of this writing, this has not come to pass.) That included a tax policy to prevent the rich from "doing what they would naturally do, which is be greedy," as one wealthy man, describing his friends and colleagues, put it.

Another approach would be taxing, through governmentally imposed "excess profits" duties, the huge profits that companies have made as a side effect of the pandemic. Excess profits taxes, which have historically been implemented by countries during wartime, impose high tax rates on "abnormal" profits made by corporations. The United States used excess profits taxes to fund both world wars, imposing rates as high as 95 percent.

Tax scholar Allison Christians writes that these excess profits taxes are "designed to counter unsustainable behaviors." You'd figure out the profits derived from, say, a company exploiting a disaster. Then you'd tax them to internalize the costs that are currently externalized—costs like, say, cities clearing snow in a climate emergency so deliveries can make it through. That would mean seeing those emergency civic costs paid for by cities and states and localities as "economic rent" that companies must pay.

Also, according to the Patriotic Millionaires and their ilk, adequate estate taxes need to be implemented. When Collins first started making this case, multimillionaires and billionaires asked him, "'Why do I have to pay this tax?' They said, 'I didn't get any government help.' They said, 'Why are you punishing successful people? Why are you burdening the wealth creators with taxes?'" After Trump was elected, less than 0.1 percent of estates in the United States pay the tax: the limit for what estates were taxed at increased from $5.49 million per person in 2017 to $11.18 million in 2018, so only 1,900 estates are now taxed under federal tax law. The fact that even large estates are not taxed means that dollars do not flow to those who need them. What if the threshold for estate taxation was lowered to, say, $3 million?

The proposal was clear. And its demotic version, "Tax the Rich" is a catchy slogan—catchy enough to be printed in faux graffiti on a couture dress worn to the Met Gala in 2021—the sane version of "Eat the Rich." As Collins put it, the Patriotic Millionaires have faith that "people like me should be taxed; that if we want a healthy society, we need to recycle opportunity; that any wealth that anyone accumulated at some point came out of a foundation of collective knowledge."

"I pay a lower tax rate than you do, which is startling," Eric Schoenberg, another one of the Patriotic Millionaires, said. To illustrate this problem, Schoenberg had posted portions of his returns online. He wanted to show how much he benefits from our system. He had always benefited from low taxation on his investment income, for instance, he noted.

"I believe today's tax proposals will cut taxes on very rich people, and that's why I posted some of my tax returns online," said Schoenberg, referring to Donald Trump's call to repeal the estate tax altogether, as well as Republican plans to cut the corporate tax rate. "Look, if you are uncomfortable about where society is and want to make a difference, this is one way you can."

I saw megarich tax hesitancy firsthand three years ago, when moderating a panel about poverty and mobility in America: one billionaire participant argued angrily against a wealth tax. (He also suggested that some housing projects be razed and their inhabitants relocated, for which he was booed.) I thought to myself, as that billionaire spoke, of how the gap between the rich and the rest of us had only widened over the years; a disturbing 2018 report found that 82 percent of wealth generated in the prior year went to the richest 1 percent, while one in five children in rich countries still lived in poverty. Given these realities, shouldn't the richest be challenged to drop their single-minded interest in individual gain and pay what they, at least ethically, owe?

———

The political efforts of the ambivalent and transparent activist rich go beyond just getting the wealthy to pay taxes. It's also how philanthropy itself tends to work. As the near-billionaire philanthropist Nick Hanauer has written, this approach can be problematic not only because of its tax benefits for the richest but also because it is an example of the fig leaf strategy that "appeals to the wealthy and powerful, because it tells us what we want to hear: that we can help restore shared prosperity without sharing our wealth or power."

Normal large charitable donations were not to smaller, esoteric groups: rather, they were to high-status institutions that might bring the giver renown and contribute further to the status and fame of the superrich. Some of these philanthropists include those I call "shock benefactors." At its worst, this kind of giving is downright egotistical.

Shock benefactors like to surprise and make the public swoon. They'll give large gifts to their alma maters, say, that tend to be already celebrated top colleges, and then get public abeyance for their generosity. In 2019, the billionaire Robert Smith was widely lauded when he announced he would pay the student debt of Morehouse College's 2019 graduating class to the tune of $34 million. Another example: Michael Bloomberg, the former New York mayor and presidential candidate, who in 2018 gave $1.8 billion to his alma mater, Johns Hopkins University. A third: Steve Tisch, who in October 2019 earmarked $10 million for middle-class students at the University of California at Los Angeles. A fourth: Ken Langone, a founder of The Home Depot, appeared on a TV show in April 2019 to celebrate his surprise gift of $100 million to help pay the tuition of medical students at New York University. Maybe it's better that they were giving their money to universities rather than sending themselves to space like others of their ilk. But these givers' efforts are also a symptom of an economic system out of whack. Their gifts don't alleviate structural problems afflicting millions but rather just palliate a teensy-weensy piece of student debt overall. In addition, gifts to their alma maters—no matter how well meaning—meant extending generosity to often already prized colleges, further evoking economic segregation.

These giant swashbuckling gifts are the ones that make headlines. That's in part because shock benefactors satisfy our narrative needs, resembling the long-lost relatives in Victorian novels who bequeath their fortunes to our heroes and heroines, making their lives bearable and who go on to rise out of poverty thanks to this benefactor. In return, the givers get a moral halo. Meanwhile, the students themselves at schools that were the recipients of these gifts didn't always feel they or their institutions were "saved" by them, especially as many had graduated into an economic system that they described to me as hostile to them. As Daniel Edwards, one Morehouse graduate from the time of Robert Smith's Morehouse debt-alleviation gift, told me, "It's a Band-Aid fix." He was happy for his "brothers who got economic freedom," but he noted that nevertheless "in every classroom you sit in, odds are one of the students to the left or right of you is still in financial struggle."

Patriotic Millionaires and Solidaire and their kind felt a change in who gets these gifts would also go a long way to altering a sclerotic system. As Solidaire, a donor group of more than three hundred highly resourced people, put it, they wished to "liberate wealth" in what has been called "radical giving." In Solidaire's instance, they direct this money to grassroots groups. So in 2021 their $7.3 million Movement Infrastructure Fund was dispensed to fifty-five such organizations with names like Law for Black Lives and Warehouse Workers for Justice. In 2020, the Patriotic Millionaires came up with a special plan that they hoped would induce greater and different kinds of pandemic giving. Collins led the charge, asking for higher donations for the next three years from all of America's richest philanthropists and not just for larger amounts but also to give these gifts to less typical and less elite causes.

These are causes far from the problematically glamorous fundraising galas and rococo philanthropic conferences with chia pudding and passion fruit spritzers.

While not a direct result, unless you were living under a screen-free rock, it was hard not to clock that the megarich like MacKenzie Scott, Jeff Bezos's ex-wife and a billionaire, were by 2020 starting to give in a more similar fashion, distributing billions of dollars more democratically, including multimillion-dollar gifts to more obscure community organizations and historically Black colleges (most educational giving still goes to name-brand universities, while only 1.5 percent of those dollars go to two-year colleges, according to the Council for Aid to Education.)

It was developments like more grassroots gifts and the recent public shifts in thinking about the wealth tax that gave Collins hope for a different kind of giving: a billionaire surtax as well as an America that is less individualistic, where paying higher taxes when you are rich is what it means to be a good citizen.

He was a member of a vanguard—he had turned down his inheritance very early, after all. And to this day, he said, he had no regrets about his own action. He remembered how his father was initially concerned about his early choice to rescind his inheritance, but said "after a day of walking and talking, he was convinced that I had not been possessed by an alien cult." It was a testament to his persuasiveness that he had gotten his own father to change his mind as well, telling his son that he had done the right thing, leaving his inheritance behind; that "the baggage" Collins would have carried if he hadn't survived on his own would have been too heavy. His kids are also OK with the descending arrow of their privilege, Collins assures me. And he, for one, is pleased that his whole family will be in a similar tax bracket. It means, he said, that unlike all too many of America's actual rich, "We now have a stake in getting rid of the system."

12

The Feeling Is Mutual Aid

A TURQUOISE FRIDGE stands at the intersection of Myrtle Avenue and Adelphi Street in Brooklyn. There are complimentary sandwiches available within, a can opener attached, and a big sign taped on the front that reads "Free Food, *Comida Gratis.*" After the Clinton Hill Fort Greene Mutual Aid group placed it there early in the pandemic, it was restocked regularly. This brightly colored Frigidaire is one of many full of fresh food that stood as sentries on street corners around the city and the country, all to relieve hunger during the pandemic. I regularly walked a mile or so from my home to those side-by-side iceboxes full of cabbages and sour cream, bearing a simple white-outlined painting of a woman and the legend "Care Is Communal."

Sometimes I walked because walking was our main activity back then but sometimes the trips felt specific. On Thanksgiving Day, before we were to eat our family meal of plenty, my daughter and I and another family went back to that fridge, dragging bags of pasta and canned goods, stocking it full. For months, New York City's residents could find food in these coolers. They could also get provisions distributed to them by Corona Courier, cyclists who brought daily bread and cash and other forms of help to those in need, one of thousands of groups organizing across the country, connected by Google Groups, Google Calendars, and calling trees. In the Bay Area, volunteers handed out free hygiene kits, electrolytes, and condoms. In Santa Cruz, a local mutual aid group raised money for county residents displaced during the California fires. In New Orleans, when the tourists stopped coming, support was organized to sustain a whole city of hospitality workers who suddenly had been laid off. In Columbia, South Carolina, Dylan Gunnels, who founded the mutual aid group the Agape Table, spoke lovingly of the "150 people in this group . . . most of them I'd never met before"; they collected gift cards for groceries, as well as for hot spots and wireless technology for indigent students. The activist Rinku Sen donated a box of vegetables and wrote an essay about it entitled "Why Today's Social Revolutions Include Kale, Medical Care and Help with Rent."

In Minnesota, linguist Anke al-Bataineh created a mutual aid network of neighbors. She told a reporter, "I thought, let's see if I can connect people who live around here, maybe make sandwiches, arrange food." Within a week, sixteen thousand people had joined the communard-sounding South Minneapolis Mutual Aid Autonomous Zone Coordination, rejecting traditional charity in favor, ostensibly, of flatter hierarchies and person-to-person help. The group also had gotten cheap eyeglasses and air mattresses to those who need them. And immediately after the 2022 leak of the Supreme Court vote to strike down *Roe v. Wade*, the 1973 decision that provided federal constitutional protections of abortion rights, thousands joined a Reddit community called the Auntie Network, where volunteers offered aid to help women obtain abortions in states where bans were likely, from transportation to lodging to care, a semi-underground lattice that helped provide funds or figured out travel. The Auntie Network and other groups like it formed the newest mutual aid front, now serving an ever more desperate population likely to have access to their reproductive needs blocked.

The mutual aid groups' approach could be marked by a levity and an air of improvisation and informality that set them apart from traditional charities. There were mutual aid online fundraisers for sex workers, who in the New York City area organized under the ancient Greek name Lysistrata (a comedy in which women withheld sex to wield power over men and end a war). Herbalists created and gave out care kits full of free healing tinctures (oat plants were supposed to relieve pandemic-induced high anxiety). People bought other people's pets kibble. Some even walked other people's dogs if the owner was ill or otherwise unable to do so. And near my apartment in Brooklyn, a volunteer group was organized by community activist Crystal Hudson, who would later become the Democratic nominee for New York's 35th District council seat, to help the aged and the financially stressed in our neighborhood.

All of these actions were described by a single catchphrase: "mutual aid." It was a rallying shorthand for help and consolation that was on the one hand radical, a set of practices associated with anarchists, and the most edgy of the self-organized left, and on the other hand gently quotidian. Representative Alexandria Ocasio-Cortez, the progressive New York Democrat, suggested both senses of the phrase and the movement that rose up around it, when she held a public conference call with an organizer of a mutual aid network early in the pandemic, and said, "We can buy into the old frameworks of, when a disaster hits, it's every person for themselves. Or we can affirmatively choose a different path."

In that year, the word *mutual*, like the word *essential*, became part of Americans' vocabulary, their word cloud during this time, a halo of satisfying, lean-on-me synonyms in circulation: requited, give-and-take, complementary, reciprocal. It was an idea that helped me and might help you: mutualism was a kind of interdependence that was closer to altruism than charity, more like societal reciprocity than simple "aid." Its aim was to enable change for the givers as well as the receivers.

Mutual aid was also part of a broader story of interconnection, countering mantras that we have read about in the previous pages: that we should ignore or deny our need for others.

I like many others latched onto the movement both because it was neighborly and because it was part of a countervailing tradition of collective action in America I had been reading about that refuted the certainty of the singular and the siren song of selfishness. While the mask-averse and millionaires swallowed up headlines during the pandemic, there were also those actively choosing civic cooperation, doing things like delivering elderly strangers' medications to their doorsteps. They were the much-needed antidote to those who preached smug autonomy and proffered rags-to-riches tales for centuries.

While the term once conjured the image of an earnest Berner carrying a Democratic Socialists of America tote bag, by 2020 it had morphed into an everyday philosophy. The mutual aid movement gained widespread traction under emergency conditions, as the economy closed, the Dow plummeted, and 9.5 million jobs vanished between February 2020 and February 2021.

This grim slenderizing explained why sometimes the line was so long snaking around my local mutual aid group's weekend food giveaway, baskets that occasionally included fresh flowers or even sachets, that it could easily be mistaken for the queue to the popular ice cream parlor on the same street. This was what a gentrifying neighborhood looked like, after all, where oxymoronic "hipsters in finance" consume existentially named Belgian beers as they pass by Hasidic matriarchs.

That group had started in my neighborhood in Brooklyn with a simple text thread created on WhatsApp "just to check in" grew swiftly until it hit the app's maximum of 265 members that spring. The group then used Google sign-up sheets and five hundred more people signed up.

Founder Crystal Hudson wanted people who had never heard the phrase *mutual aid* to know about their group and for able-bodied volunteers to be at the ready.

"Black women from the Caribbean haven't been able to buy culturally relevant foods on their fixed incomes for years," said Hudson, who is Black herself. That's why the group worked to find the people most in need—the mutual aid-ers left flyers on the floors of the few coffee shops that were still open or slid them under doors in apartment buildings. The people in the greatest need, Hudson reasoned, might not have access to a Google form. Money that was donated from more affluent neighbors was then poured into shopping expeditions and deliveries for the poor and housebound. Healthy neighbors called unhealthy ones once or twice a week. The group also had census collection drives, voter registration drives, and gathering supply drives. They held community calls about housing rights and police overreach. From the beginning, Hudson wanted to make sure that her neighbors didn't confuse mutual aid with charity and saw it instead as "neighbors helping neighbors, where eventually the tables are going to be turned, in the future."

"If someone asks us to get a six-pack of beer with their groceries, we do so," Hudson explained. "We are not here to tell people they should be eating organic or shouldn't be. If someone is cooking ribs, we buy them ribs. We don't curate."

The result was that some Caucasians in their twenties bought chicken feet and pig feet and brought them to elderly Caribbean Americans' doorsteps (by April 2021, it was reported that nationwide more than one in five Black and Latinx households didn't have enough to pay for dinner). It meant that Hudson herself heard "people cry on the phone when I asked them what they want us to buy. They told me, 'No one has ever asked me what I want to eat before.'"

Hudson had grown up in our shared neighborhood, Prospect Heights. Her early career was in sports marketing, and then she moved on to local government. She moved back into her childhood bedroom with her girlfriend so they could take care of her seventy-nine-year-old mother with Alzheimer's disease (her mother has since passed away). Hudson's Honduras-born mom had raised her on her own, working a number of jobs as a nurse, and Hudson knew she was expected to return the favor of her care.

Hudson's mutual aid helped people who were involved, sometimes in unexpected ways. One twenty-four-year-old told me that through the group he had, for the first time he could recall, "meaningful contact" with people in the neighborhood. The mutual aid work included giving away some of the local community garden's harvest to his neighbors. Before the pandemic, he had not walked into a garden since he was a little child, but when the group started to give away seedlings to help local residents grow food at home, he got involved. When I met him, he was turning a decimated lot in my neighborhood into a victory garden that grew food for neighbors in need—squash plants were already blooming amid the broken glass and ashy debris.

This was the "ethics" of "moments of crisis," he said. "We are doing what we can do," he said, "without the aid of the state." The last time I checked in with him in the spring of 2021, he had steam tables up once a week with trays of food on offer for free in a lot that was now filled with greenhouses.

———

Mutual aid was not just a response to crisis in 2020 or 2021. It was also a solution offered in the Before Times, including *way* before times.

Whether they know it or not, they are also hewing to the edicts of a Russian anarchist nobleman from over a century ago. His name was Peter Kropotkin. His 1902 book *Mutual Aid: A Force for Evolution* argues that groups that thrived the most were those that cooperated with one another. His book's claims were both exciting and severe. "The species in which peace and mutual support are the rule, prosper, while the unsociable species decay," he writes. Kropotkin was so assured of this faith that he made quite zealous claims for mutualism, among them that "practicing mutual aid is the surest means for giving each other and to all the greatest safety," and also guaranteeing "progress, bodily, intellectual and moral." That he had written this sort of thing, isolated in a prison in Clairvaux, France, and then in exile in England, apparently only added to his fervor for interdependence. That volume wasn't the end—in a 1910 essay "The Theory of Evolution and Mutual Aid," he argued that groups of organisms create traits adapted to their environment. Kropotkin's book eventually became the bible of mutualists. He had become such a touchstone that when the pandemic hit, Representative Alexandria Ocasio-Cortez cited the Russian relic on her Twitter feed, sharing one of his quotes: "We must cooperate to survive," along with the handwritten exhortation "Solidarity Not Charity!"

He didn't start out, however, as the man most likely to get twenty-first-century Americans to share flour and rolls of toilet paper. Born in 1842, he grew up in the exotic air of old monarchism. He started his career as Czar Alexander II's most beloved teen assistant. He then defied expectations for a nobleman and came to be considered a seditious activist. He spent time in Siberia. He escaped the Russian secret police. He found his true calling lecturing about altruism and mutual aid as it flowed between microbes, animals, and humans. An undeniably wild and seductive personage, the writer Oscar Wilde compared him to "a beautiful white Christ . . . [one] of the most perfect lives I have come across."

My favorite stories about how Kropotkin's embrace of mutualism unspooled in Russia's frosty taiga. On those frozen steppes, he studied fish schools and bugs and birds, and took fifty-thousand-mile journeys through the icy tundra. He observed the everyday difficulty of existing in an isolated arctic geography, as well as the generosity of indigenous societies that lived there—the Tungus, say. Domesticated reindeer carried Tungus' children. The reindeer helped them move camp and provided them with milk. For their efforts, these beasts were not seen as individual property but that of the entire clan. Because the environment was so rough, the Tungus mutually assisted one another. Due to the cold and scarcity, storehouses had to be shared ("I remember how vainly I tried to make some of my Tungus friends understand our civilization of individualism: they could not. . . ."). Kropotkin, his brain on Tungus, wrote then about "sociable species" of ducks and other animals: "The cunningest and the shrewdest are eliminated in favor of those who understand the advantages of sociable life and mutual support."

Their "collective labor" unified them and was a notable feature of human beings as well.

When I read Kropotkin on nature, I recognized something else about him: that his great influence was none other than the naturalist Charles Darwin.

How could that be? You, like me, may have learned in high school about Darwin as a thinker who conforms to the nature-is-violent cliché, the idea that the animal world is rough and unforgiving in its competition and a place of gory melodrama, where every puma or rat is a potential dominatrix. But this rendition of Darwin that many of us had been taught was not the heart of the matter. Rather, the purity of this version of Darwin's thought was the result of those who claimed most vociferously to be Darwin's inheritors.

To wit: "The animal world is about on a level of a gladiator's show," as Thomas Henry Huxley, one of Darwin's disciples, writes. "The creatures are fairly well treated, and set to fight—whereby the strongest, the swiftest, and the cunningest live to fight another day." This was some of a number of readings disseminated by biologists and philosophers who followed Darwin and then sometimes either incorrectly or exaggeratedly applied the most brutish conclusions of their master's work.

Taken in another light, however, Darwin could also be considered a proponent of mutual aid. Mutualism wasn't *just* an anathema to a thinker like Darwin. It also *came* from Darwin.

"Darwin was not an individualist," said a contemporary Darwin scholar I spoke with. Instead, Darwin bought into "interdependence," which he saw in animal and plant life "in group living, in organisms depending on each other." As Stephen Jay Gould writes of Darwin's mutualism, "If Kropotkin overemphasized mutual aid, most Darwinians in Western Europe had exaggerated competition just as strongly."

It was around the time I read these words that I came across the phenomena of the "Socialist Darwinist." These sorts of Darwinists argue the interdependence side. It turned out Eric Michael Johnson had written his dissertation on Socialist Darwinism and Kropotkin and mutual aid. (There is a whole scene of Socialist Darwinists, if you are interested.) When I spoke to the then-visiting scholar at the University of British Columbia, Johnson argued that Darwin's view of the struggle to exist is a battle where both cooperation and competition are key. Dependence and self-reliance both have a part to play.

Johnson discovered Mutual Aid Darwin half a lifetime ago, as an excited undergraduate, when he first realized that the thinker he "had been raised with—the survival of the fittest, the strongest wins"—was not the only Darwin available. Johnson directed me to Darwin's accounts of how mistletoe grew around trees, symbiotically—a homely infestation, a parasite of a sort that lives off two hundred species of shrubs and trees; to Darwin's mentions of the interdependence of bugs and birds—the bugs lived on the birds' backs. There were the ants in their farms, cheerfully laboring together; the small, homely fish cleaning the larger, more daring fish, and all the many other beasts that Darwin noticed engaging in natural cooperative behaviors that enhanced their survival. Even entities as lowly as polyps are interconnected, as Darwin wrote, as "thousands" of polyps "act by one movement" if any one of them is disturbed. Ants, birds, and mammals all share, as he writes, a desire to communicate with members of their species and others, too, "a manifestation of sociability."

Johnson, now in his mid-forties, had taken his passion for a Darwin-inflected mutualism and spent a year studying bonobos, which further supported his and what he believes to be Darwin's "pro-social" view of nature. Johnson videorecorded them for sixty hours. The famously lusty bonobos spent all of their time together with their group. If one individual bonobo walked away, another followed. "They were all about their social lives and reconciliation," he told me.

The survival of the species is not just "red in tooth and claw." It *also* depends on our forgoing individualism for something more sympathetic. As I discovered, in the long, cold days of the pandemic, Darwin, too, could be read to be arguing for "mutual sympathy." Darwin's "struggle for existence" isn't just a predator picking off its prey. It is also a gentler, snuggly-er "dependence of one being on another and including (which is more important) not only the life of the individual. . . ."

For Johnson and other researchers, this more community-minded version of Darwin guided not only their intellectual works but their belief systems. This included humanities folks who have joined their virtual institute, among them the late scholar and activist David Graeber, who wondered, "Why should worker bees kill themselves to protect their hive?" To protect their community over themselves was Graeber's answer, in his essay that also presented a Mutual Aid Darwin.

It's also a position refracted in harder sciences as well. A recent essay along these lines stridently titled "A Symbiotic View of Life: We Have Never Been Individuals." Its scientist authors view biological systems as composed of interdependent beings, rather than as a collection of hard-driving people and animals all trying to make a buck. Mushrooms, lichens, microbes, and trees are in some quarters now understood to be in a constant frame of cooperation. Oaks and fungi are BFFs.

In other words, if members of the Bootstrap Society have Ayn Rand and Horatio Alger as their lodestars, those who favor community have Kropotkin and (more or less) Darwin and a series of contemporary biologists and social theorists as well.

Those in today's mutual aid movement—ferrying bottles of common pharmaceuticals like statins by foot, car, and bike to their aged neighbors' doorsteps for free—are following in their wake and their theories now edge across the country, from Somerville, Massachusetts, to Columbia, South Carolina.

Some of the responses to the phrase "mutual aid" were from people expressing how deeply *into* these informal networks they were. "Everyone is discovering what some of us have always understood," advocate Sen wrote, her sentences taking on an almost liturgical feel. "The social ties cultivated by mutual aid are the same ties needed to fuel a historic boycott, a union organizing drive . . ." There were also social media mutual aid-ites who felt the need to express their passion for this kind of organizing and how much better they found it than traditional nonprofits. I wrote to one of these zealous fans who went by the name "Nena," apparently a college student, who tweeted pro-mutual aid and anti-nonprofit sentiments, and asked her, "Why do you feel this way?" "I volunteered for a nonprofit once," she answered.

In fact, "mutual aid" became such a ubiquitous slogan that it started to sound to some like a cliché. A mother and adjunct professor of Chicana studies put it like this, "The funny part is now all of a sudden you have upper-class families talking about mutual aid." She continued, "The term mutual aid is not radical! It's, like, last semester, dude."

By that summer, some of the mutual aid-ers were thinking of ways to expand their group's purview. For example, during the protests around George Floyd's killing, Crystal Hudson shifted the focus of our neighborhood's mutual aid group to include questions of race and policing. Hudson's dream was that New York Police Department resources would be reallocated to programs like her grocery shopping efforts.

As of this writing such dreams had not been realized. But there were other more practical positive outcomes that emanated from the mutual aid group that Hudson started in the months and years after it first began.

For instance, one of the suddenly pandemic-unemployed nannies in our neighborhood was connected to neighbors she had not known previously by the mutual aid list and started cooking for that neighbor family because the mother was about to give birth. The last I heard the caregiver was taking care of a newborn for the family.

In addition, by the end of 2020, Hudson had decided to run for local political office, to be a council member for New York City's District 35; she eventually won, becoming one of the first openly gay Black women in this role (the other was Kristin Richardson Jordan). She had been working for free for at least twenty hours per week since the pandemic began for the mutual aid group and found it no longer sustainable; it was her frustration in trying to "navigate systems that weren't designed for working families like mine" during the pandemic that convinced her to run, she tweeted, to try to create different options through formal channels.

The individualist story can erase the truth of the multiple dependencies of working families. What it ignores, furthermore, are the great metaphorical and interpersonal riches possible through mutualism, what the thinker Kristin Ross calls "communal luxury." Ross uses the phrase to encapsulate the proverbial opulence available to people engaged in collective endeavor, from the French communards to Occupy Wall Street in 2011. The mutual aid movement is clearly the latest candidate for "communal luxury." In addition, I think that there are similar riches to be found in everyday acts and relations of interdependence: the stranger who helped me find forty-watt lightbulbs on a high shelf in the hardware store; those repairing the sidewalks and roads near my house, turning broken pavement—neglect and disorder—into something smooth and pigeon-gray and reassuring.

The communal luxury Ross observes in today's activists who

work together for collective improvements can ultimately provide us with what I think of as *amplitude of the spirit*, where we may become more connected to each other. These relationships offer a different wealth than dollars, stock, or, say, nonfungible tokens.

Animals—including human animals—depend on each other in crises: mutualism is in the relationship between the monarch butterfly and the milkweed plant, where the monarch butterfly would not survive without the nourishment of the milkweed.

It's also in our need for play, for transportation, for medical care and healing, for lesson plans and literature, and, yes, for love.

For Hudson, creating her own aid group was the "silver lining" of the COVID epidemic. And it happened, she thought, because working from home meant she and so many others actually met their neighbors and interacted with them. They were not rushing off to the train, and they could "take a minute" with strangers in their midst. "Who is living next door?" she asked me, not entirely rhetorically. "Our organizations are filling the gaps, where the government has failed."

13

Boss Workers

RENEE TAYLOR WAS busy making five hundred meals when I spoke to her, as she might on a typical workday. She was following that day's menu, "tamales, tacos, rice, pico," so our conversation was accompanied by the background clamor of clattering metal serving trays.

Taylor was recruited to her current job when she answered an ad on a flyer for ChiFresh Kitchen. She thought about the job because it was advertised as "Meals on Wheels" but not just for seniors. It was prepping and delivering meals for a food service company, but it wasn't any old restaurant gig: Taylor, if she were to join, would own a piece of her workplace. Also, ChiFresh belonged not just to her but to its five worker-owners.

Taylor, who had previously bounced between various retail jobs, from barista to secretary, was nervous. While she liked the concept of ChiFresh, she wasn't sure about nonhierarchical ownership. For starters, that meant each owner-worker sharing in the decision making about, say, weekly menus.

Another layer of complexity was that ChiFresh's creator, Camille Kerr, an experienced organizer, was trying to recruit worker-owners that were, like Taylor, formerly incarcerated. Taylor had been released in 2013 after twenty-five years in various Chicago-area prisons.

Soon, however, Taylor warmed up to the opportunity. One of the elements that convinced her was not just ownership but that participating included free classes in financial fundamentals like bookkeeping, which ChiFresh sponsored. "I really want to learn everything about the business," she said. "I want to make it grow."

Worker cooperatives are businesses where workers both own and run the enterprise. The workers may also have representation on a board of directors. In addition, worker-owners tend to benefit far more directly from their co-op's economic success than they do within traditional companies, as the proceeds and the control stay with them. They are also a fine example of America's mutualist side.

Co-ops make up a small portion of US small businesses, but the pandemic and its aftermath further helped to popularize the model. According to Mo Manklang, policy director of the nonprofit US Federation of Worker Cooperatives, there are now 465 verified worker-owned co-ops in the country, up 36 percent since 2013. And about 450 more are in their start-up phase. Worker-ownership often leads to better pay, and in the case of worker cooperatives, worker wages tend to be higher than at other businesses—$19.67 per hour, according to the Democracy at Work Institute, which is more than $12 higher than federal minimum wage. (For instance, ChiFresh worker-owners paid themselves $18 per hour and were guaranteed forty-hour workweeks.) In addition, worker co-ops had a lower failure rate.

"During the downturn, people turned to worker cooperatives, typical of what people do when their government is unable to meet the moment and they've been, say, laid off," said Manklang. This was in contrast to traditional businesses. According to the Harvard University–based Economic Tracker, there were 37.5 percent fewer small businesses open in June 2021 than in January 2020.

These sorts of businesses existed against the backdrop of epic income inequality and tremendous corporate consolidation and union busting. It was no wonder that the people were drawn to a model that gave them back some power. The renewed interest in co-ops marked a return to what Emily Kawano, the codirector of the co-op development organization Wellspring Cooperative Corporation, called making a "livelihood" rather than just earning a paycheck. It was also a collective enterprise that was starting to find more political support. In 2020 and 2021, California Representative Ro Khanna publicly championed worker-owned cooperatives on a federal level. (He was a longtime supporter of the practice and had often hosted roundtable discussions on them.) At a town hall in February 2020, he said he was developing legislation supporting worker co-ops: "Worker cooperatives can be part of the solution in building the working class and the middle class in this country." Representative Alexandria Ocasio-Cortez also regularly speaks out in support of the co-op model.

In the case of ChiFresh, Kerr had "a vision": to bring on board co-owners who all "know how to cook" but whose years inside had been an obstacle to employment. "The pandemic lifted up the need for us to focus on meeting each other's needs," said Kerr. "What better way to do it than collectively owned?" And not coincidentally, during the pandemic a number of such businesses took off.

In 2021, three New York City drivers rolled out the Drivers Cooperative, a ride-hailing company and app owned and operated by cabdrivers fed up with what they saw as Uber's exploitation of their work. They say they are now the largest worker cooperative in the country, with 3,500 members at the time of this writing. As one of the founders of the group, Ken Lewis, told NPR, "Our pitch to customers will principally be to support drivers who have had a really hard time." For instance, in the co-op's early days, driver-owners did emergency food deliveries for those fearing COVID, for instance, and also drove their first customers to early voting at the behest of Ocasio-Cortez.

The Drivers Cooperative's cofounder Erik Forman, an organizer and a teacher with a taxi license, was inspired to create the co-op, he said, when frustrated drivers "kept coming to me to ask why we don't start our own app."

Uber and Lyft claim to take 20 to 25 percent off of drivers' rides, but a recent investigation revealed in some instances that the companies pocket as much as half. As we read in the account of Saori Okawa's driving life in chapter 10, that the companies take such large shares of the profits was a problem exacerbated by pandemic economic losses and risks. In contrast, the Drivers Cooperative told me they took a commission of 15 percent while also giving the drivers any additional profits as dividends.

When Forman was first organizing the co-op, he also ran online classes and encouraged drivers to discuss their experiences on Zoom with the hope of "consciousness raising," as he put it. Ninety-one percent of New York City's for-hire drivers were immigrants and often also people of color. He intended the co-op as a model, in contrast, of "rethinking ownership as a way of rethinking racial justice." Forman and Taylor are also part of what is called in some circles "the solidarity economy." What is meant by the latter is a system defined by communal as well as monetary aims.

One example of this was two worker-co-op pizza shops named A Slice of New York, located in, of all places, California. Another was how, in western North Carolina, the art of interdependence is being practiced at Opportunity Threads, a worker-owned cut-and-sew factory that launched in 2015. Opportunity Threads specialized in customizing patterns and producing small runs of products. Employees were voted in as worker-owners after a twelve- to eighteen-month vetting process. Most of the mill's staff descended from Mayans who immigrated to the region from a very-much-at-war Guatemala, and the factory's goal they defined quite loftily as building a "new working class."

And the worker co-ops were not the only co-ops that were gesturing beyond the limits of individualism with the structure of their enterprises. It was also some—though not all—of a much larger sector, the consumer co-ops. I spoke to workers at the Norwich Service Station in Vermont, and its sister outlet, the Hanover Service Station in New Hampshire. They were part of a large local cooperative that also includes three grocery stores and a community market and employs roughly four hundred people across its four grocery stores and two automotive service stations.

Located in New England's Upper Valley, a rustic spot on the border between Vermont and New Hampshire, the service stations performed basic car repairs, like fixing tires or brakes, for those in need.

And the workers at the stations told me that they were suffused with the collectivist impulse that the worker co-ops also share. Take Kegan White, a technician who completes detailed tasks like fabrication and de-rusting. For his labor, White, who was twenty-eight years old when we spoke, made a competitive $23 an hour and received dental and vision benefits, which he said were practically unheard of in the auto-repair industry. The service centers had also featured a voucher program called Co-op Car Connections, whereby the mechanics did crucial repairs for car owners unable to pay, from oil changes to fixing a broken seat belt. His favorite assignment is brake repair, as it is potentially lifesaving.

Then, when COVID alighted, workers said their cooperative took care of its people. "The grocery stores did really good in the beginning of the pandemic, and the company could have just kept all that money," White told me, "but they gave it back to the employees. I've never worked for a place that took care of their employees at that level.

"Going through the virus was an eye-opening experience for me to see how much the co-op did for their employees," White said. "They never tried to cut our hours."

A technician supervisor at the service station described the co-op as "no place I've ever worked before. I'm here every Saturday. I come in early. This is my church. This is my world. We have access day or night. Full autonomy." (If you have ever had a job, you know such positive sentiments are not commonplace.)

There were, however, some downsides to the cooperatives, including structural blocks that make building co-ops more difficult. For instance, banks tend to treat co-operatives differently from conventional businesses. The reason for this is partly the norm-busting attempt at hierarchies that is the cooperative format. "The normal business model is ownership by one or two people, with people that work for them below," said Manklang. Since a business loan for a company tended to go only to one or two people, not a dozen, co-ops can find accessing financing to be tricky. The government neglected co-ops in other ways as well: federal grants may have sustained small businesses during the pandemic but didn't do so for most worker cooperatives, as most were not deemed eligible, according to reporting by ProPublica.

But co-ops, with their model of mutualism and nonrapacious success, have arisen in other eras without widespread support. They expanded, for instance, in the 1880s, when many were founded by Black people after the Civil War. Yet another wave crested in the 1960s and '70s. How did the best ones function, and what can they teach us about community-oriented businesses today? This question led me to the economist Jessica Gordon Nembhard, who had dedicated her life to studying this.

═══

It started in childhood for Nembhard. She and her sister and two brothers grew up in a different kind of cooperative setup, a commune. The kids were comfortable talking to adults in the community, using their first names. She was one of the few people who would live, in the 1960s, in "the kind of place where there was always someone to go to demonstrations with." Skyview Acres had forty-four families, and although only four families, including Nembhard's, were Black, the stated purpose of the commune was to create an "interracial nondenominational group that lived in harmony." It was, Nembhard said, "isolated from the racism and the pressures against progressivism at public schools. You didn't have to explain why you were against segregated swimming pools."

At Skyview, everyone owned their home and the co-op controlled who was allowed to buy into the community—they also owned a shared pond and baseball field, with swimming lessons and a lifeguard, summer camps, lessons in cooking, pottery, and journalism, and Labor Day parties.

When Nembhard, then a professor, found her interest in cooperatives roughly twenty years ago, she focused within that category on Black co-ops. She was curious whether they existed or whether they were just her own wishful thinking. Was her fascination simply reflecting the cooperative of her own youth? Almost everyone she spoke to insisted that the Black co-op phenomenon didn't properly exist: "Even Black people told me, 'We don't do cooperatives: only white hippies do them.'" Nembhard, then in her mid-forties, persevered, seeking traces of these alternative businesses.

And then she found them, in the writings of W. E. B. Du Bois from the turn of the last century, who, she was surprised to discover, wrote extensively about Black co-ops. He had described them as "the first wavering step of a people toward organized social life," in his monograph about them.

She found records of idealistic groups as well as more run-of-the-mill collectives from before the Civil War. Nembhard also discovered enslaved Blacks who had shared small kitchen gardens to provide food beyond what was offered to them, or free men who started co-ops and pooled their earnings to buy other men's and women's freedom, understanding the necessity of community effort.

Later, toward the turn of the twentieth century, freed Blacks started mutual aid societies. They collected members' dues and monthly fees to pay for burials and additional stipends to widows and orphans, and they shared money and work to buy and run farms.

Their number included insurance collectives, Freemasons groups, and the predominantly Black fraternal order the Grand United Order of Odd Fellows. Du Bois himself had created a massive list of these sorts of groups, ranging from the Royal Mutual Aid Beneficial Association in Wilmington, Delaware, to the Afro-American Industrial Insurance Co. in Jacksonville, Florida; from the United Aid and Benevolent Association in Jersey City, New Jersey, to the Young Negroes' Cooperative League—a national network of cooperative groups to advance young Black people economically—led by Ella Baker. These co-ops were collective marketing projects and credit unions. To belong to them, members paid monthly dues, went to meetings, and attended cooperative community events, like dances and fundraisers.

"When we were freed after slavery and went to work, we got the worst jobs and the worst pay and the least disposable income," Nembhard said. "Cooperatives were the way to get around that."

How those early co-ops were created and conducted, and what they would mean for Blacks in the future, were the questions that drove Nembhard. If a person wanted to join a credit union, for instance, they had to be a successful businessperson with a fine credit history and money in the bank. This was a problem for people of color who were still shut out of traditional financial culture as institutional racism had for decades limited their participation. For them to get in the door of a cooperative credit union or a mutual insurance provider, they often had to know someone. But within the culture of co-ops, other kinds of collateral were accepted and valued, including trustworthiness or personal history.

Thirty or so years later, the New Deal era also encouraged a shared work ethos, and a wave of collective businesses were pleasingly named "self-help cooperatives," a truer version of the word *self-help* than what we have today, as there are no Gwyneth Paltrow–self-care–jade eggs in that version of the word.

The Self-Help Cooperative Movement started in 1931 when un-employed people in the Compton neighborhood of Los Angeles con-tacted farmers outside of the city to set up a barter system: labor for food. (The co-ops back then made more than just food, though—the worker-owned Cooperative Industries of Washington, DC, for instance, ran a laundry service, shoe repair, and cannery.) The move-ment that offered an estimated three hundred thousand people in California alone a livelihood didn't tend to make it into history books, yet the worker and consumer cooperatives became a huge movement with 1.3 million people involved.

In Gary, Indiana, in 1934, for example, when the last bank left and many were left jobless, some of the Black inhabitants created a grocery cooperative and discussion group. Around that time coopera-tives also helped spur the creation of some independent Black co-op schools.

Personal accounts of these cooperatives can be stirring. As Blos-som, a partisan of the cooperatives of the 1930s, told the author Viv-ian Gornick, "We'll remake our lives. The shoemaker will make shoes, the doctor will treat the ill, the artist will make beautiful things, and we will share among ourselves." Blossom described herself as being "thrilled" by the co-op movement's parlance and ideals, and its fun-damental rearranging of power relationships by notions of changing ownership. Co-ops were also cheery, she noted, filled with "people who cared," rather than the "gloomy" Communists.

Creating and sustaining a cooperative back then wasn't always a joyful process, and its participants could also be subjected to bias. The Cooperative Industries of DC applied three times before it got a grant, for instance. Nembhard believes there was racism at play—even a USDA-funded paper in 2002 would note that Black farmers tended not to benefit as much as their white counterparts in that time.

Yet the interest in these mutual establishments continued to come in waves. The third sprouting of the Black cooperatives included Black farmers' collectives. They included the South West Alabama

Farmers Cooperative Association and the Federation of Southern Co-operatives. More subversively, there was the mutual aid provided by the Black Panther Party, who in 1969 handed out eggs and oranges and other breakfast foods to public school children who would otherwise have gone hungry. The Panthers were demonized in some quarters as heedless militants—at the same time, getting food to thousands of children who had sometimes never eaten breakfast before, according to one newspaper account.

When she started her research, Nembhard believed putting together the history of Black cooperatives could serve as a useful model for contemporary times. Some of the power of this history of Black co-ops was surely an antidote to the racist rhetoric of individual economic success. (I think also of the words of writer Mikki Kendall, in her book *Hood Feminism*, recalling the year she was not only battling hunger but scrambling unsuccessfully to get a Christmas tree for her son, expressing how, injuriously, "Black people [must] pull themselves up by imaginary bootstraps in order to be found worthy.")

Nembhard was such an ace representative of the worker cooperative movement in part because she didn't need to be convinced of their value, after her Skyview commune-like childhood and creating a cooperative of her own, in the 1980s, with her sister. Both young mothers at the time, they ordered organic food directly in bulk for themselves and their neighbors, wholesome baby food from the warehouse, along with flour or pasta shells and cheese. It was a hobby of a sort, both quotidian and utopian. She was something of an abbess of alternative living.

Two decades after she embarked on the excavation of the Black cooperative, many of her generation and mine were still likelier to have been taught the self-sufficient ideologies of individualism in history class than the stories of collectives, inculcated back then by the tales of Great Men's singular achievements. There was a more collective-minded canon but it wasn't as widely circulated.

If we are to move away from bootstrapping, we should consider, say, the worker co-ops, and recall the words of the anarchist noble-man Peter Kropotkin, whom we met in the previous chapter: "Under any circumstances sociability is the greatest advantage in the struggle for life." (I have been saying that to my introverted husband for seventeen years.)

Like Nembhard, I had long-held dreams of greater solidarity, of a frame for the American self that looked like the brightly colored hand-drawn images in, say, Richard Scarry's books from my childhood. I thought of my favorite *What Do People Do All Day?* and doctors, sailors, foresters, firemen, house builders, et cetera—all of whom seemed, to a child at least, to own their own labor.

And during the pandemic, the cooperatives became one drawbridge out of our private castles of quarantine, to collective effort and potential transformation.

The new cooperatives were closer to a "worker-centered approach" to American culture, where interdependence is valued. They were also a refutation of the attitude that was on display in some quarters during the pandemic, where workers were abused and disdained and risked, as when seven hundred personnel were infected at the Smithfield Foods pork-processing plant in South Dakota in April 2020. The Smithfield PR flack said that the plant's "large immigrant population" were responsible for the unhealthy conditions, as "Living circumstances in certain cultures are different than they are with your traditional American family," intimating that the disease was following "certain cultures," an ugly slur, rather than the result of straight-up low pay and tight and underventilated factory conditions.

The worker cooperatives implicitly rebuked this evil and extreme version of the every-man-for-himself ideology.

=====

Owning a food service company or a ridesharing group together showed me an alternative pathway, an example of how people created their own institutions, especially when traditional institutions ex-cluded them. I also saw these co-ops as fundamental ways their own-ers, workers, and members connected with others through material shared experiences and joint missions.

These common enterprises also offered their participants a rare and affirmative response to scarcity. While some call it solidarity eco-nomics, others have dubbed it the "economy of abundance," a refram-ing that gets at the paradox of lacking in material goods but having *more than enough* collective feeling and will.

As with mutual aid groups, the strength of the collective evokes "communal luxury." At a time when we're saturated with stories of both individual achievement and individual suffering, worker co-ops can show us communal endeavor.

But you needn't have to join a worker co-op to participate in a sharing economy. Nembhard pointed out that "I lend you sugar, and you lend me sugar" while not "hard economics is economics all the same," and so is, "We carpool, and I babysit for you, and you babysit for me."

Many of us unwittingly already were and will continue to be part of such exchanges. Worker cooperatives were simply the next step. "We set our own wages and have meetings on a regular basis," said Renee Taylor, of ChiFresh Kitchen's worker-owners. "I want to make the company work for me."

14

Inequality Therapy

CHRISTINE "CISSY" WHITE is an advocate for survivors of trauma. That's partly because she is one herself.

When I first met her, she lived in Weymouth, Massachusetts, then with her teen daughter. Back then, she earned a living as a community facilitator for the nonprofit PACEs (Positive and Adverse Childhood Experiences) Connection.

Two year later, she showed me around her home: one bedroom in her house as a sanctuary, where she did downward dogs and wrote in a notebook, all to help her treat trauma symptoms, which included trouble sleeping.

The place had a very different temperature emotionally from the one she had grown up in.

White's mother—who'd had three kids by the time she was a young adult—was just a teenager when she became pregnant with White's older sister. White's father was a Vietnam War veteran who struggled with alcohol after he came back from Southeast Asia. White barely knew him but understood he was often unhoused. All of this was part of larger unhappy pattern. Before the age of ten, she had experienced the divorce of her parents as well as physical and sexual abuse. Throughout much of White's childhood, her mother was the sole provider. Growing up poor, White would hide the tape and paper clips that held her broken glasses together behind her bangs. When she went out with friends she was often starving as she was unable to afford food: to hide her embarrassment, she said she wasn't hungry. She also used paper towels as pads when she was a teenager because she wasn't provided money for those either. In that time, White and her family also moved frequently. Her family stabilized economically when her mother remarried for the third time to a relatively financially secure man who worked in what was then the digital sector. Her mother's marriage was part of how White survived poverty and neglect.

This relief from indigence, for instance, was marked by her moving from Boston's working-class neighborhood of Brighton and its Waltham suburb to the upper-middle-class town of Harvard, when she was a teenager. She said that showed her the difference between growing up in a place where the whole community is challenged by poverty and one where a single family is put upon, as hers was. White also discovered she had many more resources in her wealthier school district, for example. Due to these new splotches of social-class privilege, she learned enough about college and applications to try for plush Hampshire College in Amherst, Massachusetts. She was accepted and attended on a scholarship. College was a shocking experience for her. Her life shocked her classmates in turn. There, she was surprised to meet people who, in her eyes, *had no palpable obstacles.* "The presence or absence of adversity shapes us a lot," White said.

From the beginning, White had always prickled when she was complimented for being so "resilient," even though as the first person in her family to graduate from college she was often called that. She viewed her own story of resilience as one fortified by unexpected periods of economic freedom in her later youth and also of access to therapy, neither of which is emphasized or acknowledged in the typical bootstrapping story.

White herself reckons she grew up with a total of eight adverse childhood experiences (ACEs), from abuse to abandonment, and survived to become a counselor and an advocate. It's hard to overstate how bad eight ACEs are. With four ACEs you'd have had childhood toxic stress, because your number of ACEs are like "a cholesterol score." One gets one point for each type. The higher one's ACEs score, the higher the risk to one's health. (The understanding of ACEs starts with the original ACEs Study, the CDC-Kaiser Permanente Adverse Childhood Experiences Study.) White's job was counseling survivors in need, usually over the internet, communicating with the ACEs social network of forty-five thousand.

The main reason why she took so well to the method that recognized ACEs first and foremost, White said, is that she had a severe distaste for the idea that she—or anyone—had lifted themselves up out of pain and poverty. Her own rare success could have reaffirmed that myth of bootstrapping: her career; her capacity to help others and have a balanced family; her modest but comfortable home with its gas fireplace and a robin's-egg-blue wall and signs that offered encouragement like one that read "This Is Our Happy Place"; her put-together and pleasant appearance, with her ash blonde hair cut in a short, spikey do and her retro-cool cat's-eye glasses. But all of these achievements produced the opposite effect, as she knew too many worthy people who hadn't "gotten out" as she had. She also disliked individualism's sister, resilience. She "rages" and "recoils" when she hears either word, which happens more and more.

Instead of terms like "resilience" or "grit," White and others have embraced another solution, which her peer group embodies: "trauma-informed" community. These groups, according to White, are comprised of those who have experienced all manner of abuse and provide a place where people may congregate and seek services for free. White offered care to those who have undergone or were currently undergoing agonies like those she experienced. You might think of "trauma-informed" subcultures like White's and others as psychological mutual aid networks.

The trauma-informed approach, and even those very words, taps into a method and appellation that has become, admittedly, wildly trendy. "Survivor-centered and survivor-aware care," is what White also called it.

"It's not a 'how do we fix you'" model, as she put it. "It's 'I may have things to teach you about poverty, and you have things to tell me about navigating systems.'"

Whatever you call it, this emphasis has impacted arenas as disparate as welfare services and a university's student life, trauma-informed policing, and trauma-informed lawyering. Because it's a fashionable catchall, my friend the social worker rolls her eyes at the very phrase. A popular catchphrase can also represent real change, though: think of the proliferation and performative use of the term "the 99 percent." In mental health, the way that the trauma-informed framework alters the aperture is that people are understood as needing far more personal agency in their own treatment. Their actions are also now understood through the lens of their trauma, rather than by tracking their troubling behavior and judging it, a scolding that might well replicate the original bad feeling.

This shift in emphasis to trauma can be particularly helpful when counselors or therapists treat people with less money: so much of what tends to set them back are corrosive externalities that they then have been required to internalize. In contrast, "self-reliance" meant a particular thing to White, and it wasn't good.

In White's case, understanding her own suffering through the trauma rubric also changed her personal life: she met her last partner, Tom, through a dating site where she listed in her bio that she had many adverse childhood experiences. Emphasizing personal trauma when seeking love might not seem to be the best idea on a romantic site, but Tom, like her, was working class, with a high ACEs score. (White tested him on his ACEs on their first date.) As a hobby, Tom also did household projects like rigging up a heating system for the birdbath in front of their home, so their wild feathered friends could gather together in the chilly Boston-area weather.

The rubric "trauma informed" also changed her professional existence. White is part of a growing number of therapeutic and mental health practitioners who focus on clients' current and past economic realities. White emphasized in exchanges with her traumatized peers how she grew up working class and the way this background still wears down her older family members—an aunt, for instance, who has four part-time jobs and one full-time job.

I think of it all as "inequality therapy," a set of therapeutic practices happening around the country. It's not just some specialized new form of psychoanalysis coming out of some esoteric clinic. It's more like a framework, in peer-to-peer counseling and networks as well as counseling with individuals, sometimes at considerably reduced cost to the patients, and research by scholars examining the intersection of emotional suffering and inequity, including the study of happiness and unhappiness.

━━

If you have had terrible things happen to you at a young age, with social class injury piled on top, peer therapy that is awake to both of these elements is one answer, encouraging others who have gone through similar agitations, rather than turning to a single expert. After spending years speaking with White, I spoke to others who sought succor in comparable mental health practices. I am thinking of an adjunct professor and single mom of two who made on average $30,000 a year.

As a respite from her unstable and endless work, the woman, who lived in Portland, Oregon, entered into a peer support group. Her peers included social work professors who understood her economic frailty and the grinding professional devaluation she experienced every day. The counseling often zoomed in, she said, on how "vulnerable I felt that I was always about to be unemployed" and that "I felt like I am working 24/7 for free." She said it helped a lot.

"Inequality therapy" is small scale now, but it could become a much larger way that practitioners and clients recognize collective suffering or community recovery. (Practitioners who offer a sliding scale to patients with unstable earnings, or who are willing to accept variable payments from their patients, is a more commonplace fix.) All of these techniques would help free those who can pay limited amounts from additional shame and self-blame. That was how I found Harriet Fraad—she liked to say that the women's movement, whose second wave she partook in, should have focused also on making home life more appealing for all of us rather than just concentrating on working conditions. Fraad also recounted the story of a former patient, an abused wife, and how she had provided the woman with "a class analysis" of her home where her cruel husband ruled her and their children "in the feudal manner of the lord to a household" and the wife was in an "employee relationship" to her partner. The patient found this analysis tremendously liberating, said Fraad, for she was seeing the system she was trapped in for what it was. The tyrannized wife was led to understand that her suffering was not just a personal thing but rather "a pattern that happened all over the U S. This was what empowered her, this was how it came together for her that she was subservient and abused," Fraad said of her former client.

Fraad also observed that if a person is unhappy, their malaise could stem from, say, being forced out of their vocation by the disappearance of a legitimate economy around that work or they may feel "despair about their country as well as their social position."

To learn more about such therapeutic approaches, I spoke with Silvia Dutchevici, the founder of the Critical Therapy Center in New

York City, which leans into the realm of what I am calling inequality therapy as well. Dutchevici spoke in the lilting Romanian accent of the country where she was born and which she left before the end of its Cold War–era dictatorship.

She espoused open discussions of money and social class and said she started her sessions in an unusual way, by trying to create full portraits of her patients' incomes and resources. If a patient had little money and couldn't afford therapy, she offered a lower fee based not only on pretax earnings but also factoring in, say, money spent on children or aging relatives, she said. In that first session, Dutchevici said, she broached the sort of social or economic issues most Americans find even more taboo than sex—issues that would typically be avoided or deferred to later sessions.

"When I am making more money, I am excited to come to her with the ability to contribute more. I was very afraid at one point that if I lost my job, I wouldn't be able to pay," one Critical Therapy patient said.

"There's a belief in therapy and in America that 'if I am not successful enough, it's my fault,'" said Dutchevici, and "that 'I am not working hard enough on my own self-esteem.' This is wrong. Emotionally and unconsciously, people believe that if they are poor, it's their fault—therapists need to address this." Another relevant element is how much patients pay for counseling, as most therapy sessions are not covered by health insurance to begin with. Dutchevici said she tried to talk honestly about "access to mental health and if you have shitty insurance, for example, you will not get what you deserve."

This candor can feel unexpected and surprise patients. One of Dutchevici's clients was a father of four who made his living working with children on the autism spectrum. He said he felt there were certain topics "a therapist won't touch," like money, social class, or power, "the dynamics in society." The man was devoutly religious, yet he felt he couldn't share the questions he was beginning to have with others in his Orthodox Jewish community. He had to "put another veneer of narrative on the experience," as he put it, in spite of the reality that "we all deal with money and power on a daily basis."

This time in therapy, he said, he experienced just the opposite. Money and class were some of the sessions' themes.

"The goal of therapy is not just analysis but also the adaptation of the individual to an oppressive system," said Dutchevici. The latter was a negative for her and others who practice as she does.

I was eager to find context for these therapists' outlooks. So I plunged into some of the academic research around a more collective and class-based emotional view. I spoke with Dr. Kalman Glantz, something of a pioneer of class-based therapy, a psychiatrist with a distaste for the rhetoric of grit and pluck. In his Boston practice, he had attempted to detangle some of the ways in which our market-driven society distorts self-perception, filling people with a sense of failure if they rely on the assistance of others.

In his book *Self-Evaluation and Psychotherapy in the Market System,* Glantz and his coauthor J. Gary Bernhard examined how the competitive nature of today's market system may be at the root of many clients' problems or lack of self-esteem, one that forces them to constantly reassess their status. What is valued, said Glantz (with an emphasis on the commercialist double meaning of the word *value*), is "achievement rather than belonging," and his practice keeps this in mind. From the viewpoint of Glantz and Bernhard, economic inequality changes how we feel about ourselves and each other. "Self-evaluation—the struggle to achieve a high opinion of self—[is] exacerbated by the market system," he writes, which leads to stress and self-involvement.

A society as individualistic and economically punishing as our own can degrade people psychologically, so after I spoke to Glantz, I started to look at other studies that explained some of how these depredations worked. One was a 2020 analysis of US adults ages thirty and over, where J. M. Twenge and A. B. Cooper looked at the correlation between income and happiness over a forty-four-year period and found that the relationship only intensified over time. (A previous well-known 2010 study by Daniel Kahneman and Angus Deaton demonstrated that, for Americans, higher salaries were associated with increases in day-to-day satisfaction, with well-being tapering off at an income of about $75,000.) In another article, "Income Inequality and Depression," the authors, led by Vikram Patel of the Department of Global Health and Social Medicine at Harvard Medical School, conclude that "nearly two-thirds of all studies and five out of six longitudinal studies reported a statistically significant positive relationship between income inequality and risk of depression."

I was also struck by some of the research I uncovered about how some therapists treated patients who were financially stressed. According to one study entitled "What's (Not) Wrong with Low-Income Marriages," therapists tend to query poorer patients about their attitudes toward marriage, as if they were already in some way suspect, a questioning that stems from stereotypes. In truth, the poorest Americans have more traditional views on romantic unions. They might have even longer marriages than the affluent if their economic struggles didn't haunt and tear them apart, a fact that helps to account for poorer couples' higher divorce rate.

Also in the class bias annals was a 2020 analysis by Princeton University, in which researchers found that the poor are perceived to be "hardened" by their struggle and thus less wounded by it than those of means. They called this a "thick-skin bias," and the researchers believed that this was partly to blame for why so many Americans don't clock the distress of the poor in the first place. In addition, therapeutic ideas of "dependence" have for decades contained a bias, like not recognizing that some of the people considered emotionally dependent appear so simply because they don't have the financial resources to change their lives and get out of what are sometimes called "co-dependent" relationships. (The concept of emotional dependency doesn't always "recognize or confront the social and economic realities in people's lives," as social psychologist Carol Tavris writes.)

Finally, getting therapy at all is also much harder if you are lower income. This was the experience of the writer Katie Prout, who, in an essay my organization commissioned entitled "Medicaid Has Been Good to My Body, but It Has Abandoned My Brain," explained her search for professional help covered by the program while living in Illinois, one of the "designated mental health shortage areas." She chronicles how she wound up being given the slip by one therapist who had contacted her after her calls and emails. "I hadn't been ghosted since my twenties," she observed, "and certainly never by a mental health professional."

Her experience chimes with a survey by the American Sociological Association's *Journal of Health and Social Behavior*, which found that New York–area psychotherapists were less likely to offer appointments to those callers they *perceived* to be working-class or poor callers, whatever their race: 8 percent were offered an appointment, even when they had competitive insurance policies.

Reading these findings, I think back to my own youth and childhood and how my economic status and mental health were intertwined. By the time I was a young adult, I had perhaps foolishly found what I thought to be respite from the culture of the self-made in poetry, a genre where the only paper is within monographs rather than currency. (Horatio Alger Awards are not given to poets.) I was trying to ward off the vampire of survivalism.

I saw a therapist and was paying her—even on a sliding scale—more than I could afford. I told her I needed more work to be able to continue seeing her, in her office filled with wooden sculptures. She replied twice that what I meant by "work" was a need to "do the work." The therapist was referring to the work of exploring my inner mental life. Looking back, I want to shout, "No, I really just need a job."

I learned Cissy White had been diagnosed with ovarian cancer a year after I first met her. It's a disease that is often resistant to treatment. She said her cancer had been responding well to treatment, however, even though it was detected later than she would have wished.

"Even with chemo and cancer and going through the COVID pandemic while recovering, I still feel privileged," White said, with typical fortitude, a Boston clip to her voice giving it that extra bit of flintiness. "I have health insurance, my daughter had insurance, and we have a computer for online schooling."

Although many might backslide to embracing ideas like resilience while enduring a difficult disease, White refused. She bore her treatment with grace. She even found it even more offensive to be called self-reliant and resilient now that she was sick.

White continued to question extolling personal capacities to triumph over adversity when those roadblocks are created by institutional scarcities. It reminded me of writing by Joanne Jordan, a senior research consultant in climate resilience at the University of Manchester, that "resilience is increasingly becoming the new buzzword," but "as commonly understood, [it] is inadequate for understanding the intersecting vulnerabilities that women face." Using words like *resilience*, by these lights, meant denying the structural forces that shape who is most at danger. Being less empowered or less affluent by birth, for instance, is at least as important a risk factor as an individual's capacity to endure setbacks.

White hated the idea that her survival was showing her to be a more successful person than those who perished. No, she said, she might just be one of the lucky ones. She and I chatted over our screens, and she showed me sculptures in her house. Some were around the theme of trauma. One consisted of twenty-five pairs of old eyeglasses and sunglasses: pairs were intact or were broken, mangled, and missing arms. White said the idea behind this work was to show what it is like to see through trauma, a lens that can distort, mangle, or intensify.

While grateful for her access to the resources she needed to treat her illness and guide her daughter through distance learning—not to mention all the food and get-well gift cards dropped off at her home, missives from the community surrounding her—she didn't attribute her survival to personal grit. What White called her "class switching," moving as she had from so poor to middle class, made her bracingly aware that she was no better as a person before or after she became relatively privileged.

She'd led webinars during the pandemic, with ACEs community members joining on Zoom, their faces lined up in rows. White's smile was bright, and an arty scarf was wrapped around her hair, still cropped from chemo. One woman spoke of her own abject childhood and young adult poverty, how she once spent her days "drinking so much I was dying." White nodded in acceptance. A chorus of attendees also responded.

The events of her life might have caused bitterness in someone else—the fact that her cancer was not found earlier even though she visited the doctor multiple times and had a tumor the size of a grapefruit; the memory of her mother being so poor that she had gotten arrested in front of White, a child at the time, for bouncing a check (the police came to the house and brought her to jail in her pajamas). White, in contrast, didn't retract. Rather, she seemed almost rigorous in her generosity.

She joined an ovarian cancer survivor advocacy group, whose members wrote down their experience of the disease for medical professionals to read, sometimes feeling as if they were testifying to their suffering and educating the "professionals," doing so against their own ticking clocks. She hung out with her partner (eventually they broke up from the strain of it all). She treated her physical symptoms with as much levity as possible, literally by bouncing on a trampoline, because that helped her treatment-swollen legs.

Months later, when I spoke to White again, she had just received more bad news. Her ovarian cancer had recurred. "All hopes for 'the cure' are now gone," she writes to me. "I'm not without hope or a fight and I will be living my healthiest and best life as long as I can."

Still later, when I last spoke with White, in 2022, she was, at fifty-five, taking part in trials of new cancer drugs. Though thankful to get to be a "guinea pig"—and to still be alive, as she said—the drugs were taking a toll on her. She had had a temperature of 100.5 degrees for weeks and didn't have the highest hopes for her immune system. Nevertheless, she spoke about the results of her participation in perhaps helping other women with ovarian cancer across time, which seemed to me to be a kind of transhistorical mutual aid and very much in keeping with how she had lived her life.

"I don't want to be part of a clinical trial unless it's disrupting the model," she said, and "pushing things forward even if I am not in the group that benefits."

Needing others, and even depending on them virtually, had helped White through the trickiest hump in her cancer. Saying that we should pull ourselves up by our bootstraps and show more resilience is an example of "well-disguised victim blaming," said White. "A lot of what we call 'resilient' is really what you might also call 'well resourced.'"

15

Volunteering Ourselves

AFTER HIS WORKDAY ended, Dr. Armen Henderson donned a respirator mask and gloves and began his second shift, his evenings spent administering COVID-19 tests to Miami's homeless. This was early in the pandemic, when such things were hard to come by, and he had to get the kits donated to his cause by a local pediatrician and CEO. Henderson devoted twenty hours a week to this treacherous unpaid work, which included both the testing and the full-service shower site his group ran. He described this labor as "civil disobedience." Henderson, whose day job is as an assistant professor of medicine at the University of Miami, and the team of fifty volunteers in the group were, among other things, disobeying the Miami–Dade County order to stay at home. In addition, he and his organization gave out hundreds of free COVID tests as well as tents to those experiencing homelessness.

As a Black doctor, he felt honor bound to treat another population that was often historically neglected or poorly treated by the medical profession. As the novelist Zadie Smith wrote in an essay about American medical care in 2020: "We had 'unequal health outcomes,'" where a public health emergency was considered a private emergency and a personal problem. Or, as Smith puts it: "Wrong place, wrong time. Wrong skin color. Wrong side of the tracks. Wrong zip code . . . Wrong health insurance—or none." This was what Henderson was seeing on the streets.

Henderson and his collaborators were part of a network of nonprofits and groups where governmental or city services were performed ad hoc by citizens, volunteer activities that carried a fresh valence of urgency during the pandemic.

For instance, some of those whom Henderson and his colleagues treated went to the hospital by ambulance after the volunteers took a look at them, or the volunteers drove them to the hospital themselves, at personal risk. (He gained unrelated attention when a Miami police officer handcuffed and detained him in a dispute, which Henderson said was indicative of racial bias.)

The pandemic isolated many, but it also encouraged novel forms of voluntarism to take root. This ranged from medical volunteers like Henderson to ordinary citizens joining in their local government meetings through something called participatory budgeting, a process of civic engagement where members choose how to spend part of a public budget for their community. In both of these instances, they were practicing an acute version of what I have called the art of interdependence. It included ragtag bands of vaccination volunteers who tried to help get needles into arms: in 2021, it meant getting the "unvaccinated but willing" 10 percent of the American population to venues to get their shots.

As with the dystopian social safety net you read about in chapter 8, this voluntarism wasn't wholly good, although it was sometimes more overt in its aims and in the societal shortfalls it meant to address. It, too, was a symptom of our troubled American systems, while at the same time being a means for people to band together, asserting their agency. This paradox was overt in the early days of the COVID epidemic. In April 2020, in addition to asking for donations of respirator masks, the Boston Medical Center sought out volunteers who could provide support and services to unsheltered people who were isolated and quarantined. Across the country, psychologists and social workers offered free tele-counseling to the distressed. Thousands of people were designing and sewing face masks in their homes to give to medical workers. Then again, why was richest country in the world unable to source protective equipment for many health-care workers and instead relying on the efforts of overworked volunteer doctors like Henderson and guys with Maker Labs in their garages? In 2020, I spoke to one on the phone who explained to me that he spent his free hours creating medical shields to be dispersed for free but he usually had to wait till after dinner to make them.

Henderson, for one, was angry about it all—the word *engagement* is one letter away from *enragement*—about the years in which things could have been done for them and their kin but weren't. He had started a local organization long before the pandemic, right after Hurricane Dorian—the Dade County Street Response—to support financially stressed communities through disaster relief efforts. South Florida had been "hit hard by climate change and flooding," and, as he put it to me, "The attitude of people living here, old people and poor people, is that they don't understand the magnitude of the threat and also they don't have the money or the means to get out of harm's way." His volunteer disaster-recovery work during hurricane season included creating "pop-up shelters" throughout Miami for those who can't afford to leave their homes. Henderson's pack was a small outfit: their budget when we last spoke was $10,000 a year. But even on that shoestring, his team found infectious individuals living on the street. They had convinced some of the patients to go to the hospital for treatment. "We shouldn't exist," said Henderson. "It shouldn't be us volunteers. But we do what we have to do as the government forgets about poor people."

Henderson grew up in a similarly overlooked pocket of Philadelphia. His parents' combined income of around $35,000 supported a family of six, including "four kids in dilapidated housing where factories spewed chemicals into the pipes and schools closed because of asbestos."

His life story showed him, as he put it, "that capitalism has ruined poor people across the country. This is why this is my calling."

Henderson had an autobiographical investment in these community-building practices. But not all the volunteers did. Tricia Pendergrast, who was twenty-nine, was one of the leaders of a medical voluntarist group that emerged during COVID, Get Us PPE Chicago. At the height of the pandemic and its supply chain ruptures, the national Get Us PPE solicited donations of masks and other medical necessities and distributed them to hospitals. That local Chicago branch was mostly composed of medical students—some five hundred strong at any given time since it started with roughly two hundred people. Pendergrast was a first-year student at the Feinberg School of Medicine at Northwestern University when we spoke. She helped crowdsource donations. With the money, her chapter bought coverings that protect the top of health-care workers' heads from infectious particles as well as N95-equivalent respirator masks. It all took a substantial effort partly due to the astronomical so-called COVID surcharge of paying for shipping to get the masks from China to the United States—$13,000 over what it would normally cost to fly the masks to the United States, because her group didn't want to wait the several months for masks to get here by sea. Time was, after all, of the essence. Pendergrast was friends with young health workers "writing living wills and sending banking information to spouses because their work [was] now so incredibly dangerous," she said. "I couldn't take it lying down."

Like Henderson, Pendergrast dove into her voluntary work, setting off for local hospitals, delivering donated gowns, gloves, and handmade masks sewn by a crafty posse of Northwestern Law School and University of Chicago students and local high school students, as well as other citizens' sewing groups.

Even in the heart of the worst of COVID times, these volunteers also admitted to feeling ambiguity about what they felt was being required of them. As one of the medical students who worked for the protective gear–making volunteer group told me, "I wake up every day and hope today is the day that [the group] won't exist. I hope that day is coming very soon."

Volunteers like Henderson and his helpers should not be our first and sometimes *our only* response to other citizens' suffering. Nina Eliasoph, a scholar of voluntarism in this country at University of Southern California at Santa Cruz and the author of the book *Making Volunteers*, said that while we think we are engaging in countercultural or collective action when we volunteer, our volunteerism also has a dark side. It gives our governments an excuse not to take care of their citizens. Historically, "the idea was that getting the federal government involved would undermine community spirit and would degrade the recipients' morality," so voluntarism allowed those in need to be helped while not requiring anything from official legislative channels. No matter what the natural or social disaster was, from floods to the Dust Bowl, Eliasoph said, Americans have looked to volunteers and not to their legislators. And within voluntarism itself, there were efforts to distinguish between bootstraps-worthy help and simple charity. Clara Barton made this distinction when she founded the American Red Cross in the 1880s, expressing Americans' emotional makeup when she said charity made people dependent while, as Eliasoph put it, "citizen-to-citizen aid was, to her mind, more respectful." Barton reframed charity so that it was shorn of its altruism when she famously said she was "offering a hand up, not a handout," implying that handouts to the poor were a source of shame—not noble like a hand up, and that it was more dignified to receive aid from peers.

The common wisdom over the years was that voluntarism functioned like a seesaw. This faulty notion, that if government aid goes up, the need for volunteerism goes down, has been disproved globally: some of the countries with the world's highest volunteerism rates, like the Netherlands, Denmark, Canada, and New Zealand, also have strong government-supported social safety nets. America's rate of volunteerism is by some measures the highest of all. We may be big-hearted. But it's also because when disaster strikes we too often have no other recourse. If we look at the Danish, with their volunteers *and* their substantial social net, we can see it doesn't have to be that way.

======

Searching for a different way of existing in our bootstraps society led me to talk to dozens of people devoted to collective efforts, from mutual aid to voluntarism. And it was through these conversations that I encountered another form of communal engagement that at first seemed highly unglamorous. Even its name, *participatory budgeting,* evoked humdrum, earnest speeches and airless rooms. But there's a lot more than that to what those in the know call PB. PB had even been dubbed "revolutionary civics." Because of PB groups—bands of citizens investing their time in their very local governments—there has been an increased action around neighborhood concerns: park spaces, say, or paths creating accessibility to the beach for the disabled, or, as one PB attendee said, money spent in ways that were not how "government money was usually spent." A chosen citizen facilitator, a sort of a tummler of whiteboards, or an emcee revving up those gathered to take part in public brainstorming, may start the meeting. The participants applied dot stickers or raised hands to show support for one citizen measure or another. The small groups were composed of people who had never previously publicly aired out their civic wishes. Now, they were discussing everyday dreams—a water element in a local playground, for instance, during hot summers when kids cry out for sprinklers, or costly curb extensions to be built on tricky crosswalks at busy intersections—could actually be realized. In

2021, a ballot included the request to translate tenants' rights handbooks into Urdu and Uzbek. These proposals were then voted on by the neighborhoods' inhabitants at city council offices or schools or even a pop-up table in front of the local grocery store.

PB was one of the small but substantial community solutions that emerged when we put our mania for self-sufficiency to the side. But PB meetings were not just quotidian discussions that led to quaint proposals for new murals or better streetlights. In the era in which police abolition was no longer a startling notion, PB had also become a more substantive theater for change. Participants pressed in meetings for what activists called "budget justice," which meant things like reallocating monies that once reliably were funneled to a city's police force.

The number of people taking part in PB processes are, as of this writing, roughly 150,000, since it was imported to the United States ten years ago. In 2020, it led thirty-three New York City Council members to ask residents to consider how to distribute $35 million in funding for improvements to parks, libraries, schools, public housing, and streets. Each district that participated had to commit at least a million dollars to the process. When PB first started in the United States, leaders of the nascent movement told me the meetings were also comprised of participants who had never felt part of their local governments and didn't know how they worked. Sometimes they felt they were not part of their communities, either. But through the meetings, they began to feel like they were "in the know" and to have more of a sense of both communality and ownership. Yet these everyday people had a positive personal quality that the government officials at the meetings tended to lack, in particular, intensity.

It was around the time that the Occupy Wall Street movement began that PB also caught hold, an after-the-recession moment of distributed social protest. One of its early American pollinators was a nonprofit worker named Joshua Lerner, among others, who imported the concept and founded one of the first PB organizations, People

Powered, that helped launch PBs in New York City, Chicago, and Boston. (Lerner had first learned about PB while living in Argentina when tenants in some of the housing projects there used it as a strategy and explained that PB had been practiced in Brazil since the late 1980s and then in thousands of other cities.) And as of this writing, an estimated 436 local American PBs have been set up around the country, in places like Boston; Vallejo, California; and Greensboro, North Carolina, as well as "People's Budget" campaigns in places like Nashville.

The PB procedure was also spread by people like Shari Davis, the executive director of the Participatory Budgeting Project (PBP), in the Bay Area, an organization that teaches the PB processes to school groups and philanthropies, among other institutions. Davis had developed an affection for participation in the broad sense when they were a child (Davis's pronouns are *they* and *them*). It was then that they realized that there was something wrong with how poorer people of color, including members of their immediate family, didn't feel connected to their municipalities. Davis's PB advocacy group engages with such groups in roughly fifty cities, concerned with what they called "collective power and sharing power."

So what does shared power look like in real time?

I am most familiar with PB in New York City, where I live. Here there are many yearly meetings in each voting district of the city's boroughs as the idea behind participatory budgeting is a simple maxim: the truth of government is its budget. And by getting ordinary citizens to make direct decisions on their local governments, the idea is that PB has made politics not only more open but also more interdependent.

A city council person might throw a meeting for their district's citizens: the most committed spend a few hundred dollars to attract participants, with flyers targeting local groups and congregations. The crowd, fifty to a hundred strong, will gather in an indoor public area, like a school cafeteria or a basketball court. Many of the attendees may have never been to a community board meeting. By day, these PB attendees may be a secretary at a dentist's office or a para at a school. They were also not usually civic "regulars" and tended to be strangers to one another.

At the best PB meetings, there is food on hand for participants—fried rice or bagels, perhaps—and there's a children's corner, an on-the-fly version of childcare, with the din of neighborhood kids acting as background noise for every meeting. There may be interpreters available or printed maps of city council districts, as often the lines of the district are not intuitive, even to those who have lived there since birth.

At one meeting, people clapped if they liked the way money from their city's government was spent in their neighborhood. At another, the answer to this question was a deafening silence, made all the more noticeable by the setting, a large school basketball court. The facilitators sorted the folks in attendance into smaller groups, by either interest or demographics.

The ages of those in attendance varies, but some of the most ardent participants were startlingly young: when City University of New York political scientist Celina Su surveyed twenty-five thousand PB participants for her research study, she was pleasantly surprised to discover that children as young as eleven wanted to participate in PB. In one public high school where students felt physically threatened, the students used PB to talk about the conditions in which they might feel safe on campus, where some students might not fear the aggression of other students. At another participatory budget meeting, students said they needed better rooms to hang out in and to get to know each other better, so they could better detonate any fights or violence that broke out. This led to the school renovating common spaces, connective and pacific places where students could congregate. PB spurred the installation of working washers and dryers for seniors in one public-housing building where the existing ones were broken. It also brought about the creation of an eighteen-inch bronze statue of the beloved stuffed rabbit from a children's book—*Knuffle Bunny*—in the garden of a library in Park Slope, Brooklyn. It was built after the project won New York City comptroller Brad Lander's participatory budgeting funds in 2015: it was sculpted in a neighborhood artist's apartment. Also, in that same tree-lined, privileged Brooklyn enclave, funds had been given to repair a public school's deteriorating bathrooms, because few kids were able to find working facilities during the school day. As the school principal said at the time, "I've had students tell me 'no more holding it in.'"

Participatory budgeting also sometimes offered citizens a forum for expressing their discontent with their local police forces, by withholding or attempting to withhold local funds from them. (This has become a more common realization: in New York City, for every dollar spent on police, only one cent was spent on the general workforce, according to Alexander Kolokotronis, a young scholar who specializes in participatory budgeting.) In Nashville, in the summer of 2020, PB advocates demanded that the police budget be reduced and social services be more adequately supported. While that city's Democratic government rejected the PB advocates' demands, a movement was born. Similarly, in Seattle in December 2020, the metropolis's city council voted on whether to cut the police budget. There were four hours of public hearings, with two hundred callers signed up to speak. Many PB participants wanted to raise taxes on Seattle's large businesses. In the end, the budget was cut by 18 percent, with $30 million put in a fund that will be distributed, not straight into social services but by a participatory budgeting processes, toward those experiencing homelessness.

The process turned PB attendees' "attention from their plight as individuals to the conditions of their community," said Celina Su. Sure, there were the neighborly discussions about a bench shortage or how to protect local park visitors from a bat infestation (both of these examples are real). But the process was also, in Su's view, "interdependence itself," as it engaged many thousands in improving a place's conditions, even if those details don't affect them directly. For Shari Davis, the PB advocate who had, as they put it, grown up "really low income," the power of PB was that it encouraged those who were disenfranchised to believe their local governments.

One name for this new faith and the engagement PB brought about has been called "civic literacy." It's a phrase used by the media studies scholar Henry Giroux to describe how we learn to "read" and operate productively within existing structures of power. The participatory budgeteers and the volunteers, including Dr. Henderson, reminded me that, despite its aggravations, civic literacy and societal generosity can help those who give as well as those who receive. When we open our proverbial front doors to let others in, as the volunteers and the PB enthusiasts do, we may find ourselves relieved of our own melancholy for a moment.

In the mutualists and the participants, the peer therapists and even the transparent rich, we saw people interacting with the publics around them.

Such commitment to the commonweal can be taken, in some quarters, as an affront to the American religion of individualism. It can even provoke fear that the *self itself* will be whittled away or engulfed as a result. These are trepidations that we must push away, though, so our liberation into a more collective existence can begin.

16

Unmaking the Self-Made Myth

THE PANDEMIC ACTED like a proverbial X-ray machine, exposing how self-defeating our isolation from one another can be. The pandemic also exposed, with the clarity of medical imaging, how dependent we are on one another. (The X-ray metaphor is a cliché but, like many clichés, is also true.)

For me, the pandemic also revealed yet again how the self-made myth underpinned the fable of the American Dream. The fiction is that we must strive on our own for singular success that is within reach of us all. This notion doesn't just provoke foolish materialism or inevitable petty disappointment. It has had serious negative effects on our social fabric, sustaining inequality and hindering the better collective choices we could be making.

But we shouldn't just linger in this failed dream. Let's consider how to get a new American Dream, if such a thing is possible at all.

There are small-scale, tangible things we can start with. Mutual aid groups, cooperatives, participatory budgeting groups, community trusts, and the like help launch us on our voyage away from the trap of individualism.

Pulling ourselves up by our bootstraps has hurt us: we need to instead forward accounts that oppose the shaming of our country's most vulnerable. Part of how to do this is for us to redefine this contempt toward the financially stressed for what it is, societal bullying.

Countering these poor-bullying story lines starts at the top. We need counterprogramming if we are to reject our fake just-folks political leaders who are cosseted wastrels pretending to be self-made. Politicians should instead tear up their usual Horatio Alger spiels and, as counterprogramming, find appealing new ways to say the equivalent of "it takes a village."

Some examples: in 2019, Robert Reich, the labor secretary for former President Bill Clinton, offered "a casual reminder that 60 percent of all wealth in America is inherited. 'Pull yourself up by the bootstraps' is a sham." In 2020, AOC remarked during a House of Representatives committee meeting that "it's a physical impossibility to lift yourself up by a bootstrap, by your shoelaces," and continued to say "the whole thing is a joke."

Leaders might also emphasize their vulnerabilities, recounting moments of interdependence and dependence in public forums. This would make both interrelatedness and fragility more socially acceptable. In addition, if we are to battle the age-old story of bootstrapping success, peddled by everyone from the prairie propagandist Laura Ingalls Wilder to Ronald Reagan to representative and gun-rights obsessive Lauren Boebert, we need to deploy fresh hybrid narratives. What I mean by this comes by way of articles, films, TV shows, and so on that balance the psychological with the sociological; that are at once firsthand and human and structural and expert.

An example of this technique: when the character on the popular television show about a British soccer team, *Ted Lasso*, the kooky

Coach Beard, says, "You know, we used to believe that trees competed with each other for light. Suzanne Simard's field work challenged that perception, and we now realize that the forest is a socialist community." In this aside, Coach Beard makes an argument for interdependence, about how trees don't compete with each other for light and instead "work in harmony to share the sunlight." It's a joke and an insight, based on the plot of a goofy TV show, where the moral is always teamwork is best, but crossed with cutting-edge ecology that supports teamwork and community as an essential part of nature. Storytelling like this that mixes abstract understandings with concrete, everyday details and quips has been found to be one of the best tools for shifting public perception, including altering the widespread demonization of the indigent.

Further along those lines, lawmakers and school boards and teachers who spent years on "grit" curricula could instead launch national school initiatives around the value of interdependence—of being together and leaning on others. As of this writing, the very opposite was happening. But what if we could try to bring into the schools Cissy White's trauma-informed outlook or the mutual aid movement or forest ecologist Suzanne Simard's theories that our arboreal friends communicate through belowground fungal networks, led by hub mother trees, entwined for their survival, as quoted in the show *Ted Lasso*? All of these frameworks contain valuable and exciting lessons. However, in order to propagate them and teach kids something inspiring and blame-free, we need to actively recruit community-minded people to schools' ranks.

Another way we can work toward greater interdependence is by building off of the most beneficial political attainments of the last few years. For instance, in 2021, the Biden administration's American Rescue Plan has shown us what the government can do for families in an emergency. The plan, among other things, offers working parents a significantly higher child tax credit, up to $3,600 per child, with an additional $15 billion to help low-income families access childcare and to cover COVID leave. It was a progressive and necessary step, perhaps the most left-leaning economic plan since the Great Society.

Even though the Biden administration's attempt to offer adequate assistance was often blocked politically, the very existence of such a proposed set of supports seemed to promise transformation. And the changes that were adopted showed the necessity of collective solutions, how we can work together to support our most at-risk and underwater members when they are imperiled. As of this writing, it was mostly just temporary. Making these shifts permanent is what's needed next: a genuine universal childcare system with high-quality facilities and staffing and solid salaries across the board would be a yearly $70 billion annual proposition. This is the sort of investment we now require, not just this year but every year.

Part of how we can achieve making this sort of money sustainable in the long term is additional tax dollars. This could be derived from the wealth tax or a far lower threshold for estate taxes than the $20 million per couple that currently applies. Another alluring notion is that we raise money to pay for some of this infrastructure by "excess profits" taxes—where profits made by corporations above an 8 percent rate of return on their capital are taxed at progressive rates, as we have done in other moments of disaster and war. After all, by August 2021, America's billionaires could have covered a $3,400 check to all 330 million–plus Americans and still be richer than they were at the start of the pandemic: they had had a wealth gain of $1.8 trillion. Just let those numbers wash over you.

In the meantime, Big Philanthropy—that collective of high-net-worth individuals and foundations—should further democratize their giving. The greater need for this is something I have experienced personally, as someone running a compact nonprofit that was parceling out grants during the pandemic's early months. Back then, it was all about supporting writers and photographers when they needed it and not making them "pull themselves up" when they were down. Our group wasn't alone in our more direct efforts, though: we talked with members of other organizations who were also giving COVID relief funds to artists and journalists.

In my own pandemic workdays, carried out at the kitchen table, I devoted myself to helping at a distance, and raised and distributed emergency financial aid to my reporter colleagues. *Give people money when they need it, without elaborate rings to jump through*, I'd mutter to myself, as I worked on behalf of struggling journalists in an epoch that was called a "media extinction event." Certain grants were not intended for emergencies but for cultural production. Still, when an estimated twenty-seven thousand media workers lost their jobs during the pandemic, and many publications froze or eliminated freelance budgets, threatening the livelihoods of independent journalists, reporters had to have funds quick. An example: a photojournalist wrote to me in the spring of 2020 in desperation, as they had COVID and no savings. They were only able to pay two-thirds of their monthly rent, thus fearing eviction. Our organization sent them a check and didn't ask for the usual photo essay in return, a kind of "disaster giving."

I wondered whether the way that we started to parcel out money in those months should become a model for future giving. As Sofía Gallisá Muriente, a thirty-something artist who had co-directed the nonprofit Beta-Local, based in Puerto Rico, told me, she wanted such reactive disaster funding to continue in ordinary times. As she put it, "People need to be supported in ways that value their lives over their projects." At the very least, the nonprofit sector should try our best to not act like controlling spouses, reviewing our grantees' spending with magnifying glasses. Rather, as much as possible, we should offer unrestricted funds.

My fear of what could go wrong if philanthropies and nonprofits were *too* careful and parsimonious in their funding of individuals was encapsulated by the words of the artist Alice Neel, which accompanied one of her canvases at a show of her paintings at the Metropolitan Museum of Art during the pandemic. The 1933 painting depicted a day at a poverty foundation, wrote Neel, where "the woman seated in a chair in the middle of the picture "was living with her seven children under an overturned automobile—that was their house." The foundation "never gave a penny to the poor" but simply "investigated the poor. Out of that came social security and welfare, but before that, you just starved to death." Reading Neel's words only added to my conviction that nonprofits and philanthropies should—as much as possible—not just interpret but, especially in a time of economic emergency, *do*. In a sense, I thought, we were now always in a disaster, a result, in part, of shock capitalism.

———

At this juncture, you may be thinking: OK, thanks for all your helpful hints! After all, I assume that you, dear reader, are not a politician with an unusual amount of societal power, and there's a good chance that you are also not wealthy enough to change the course of taxation and philanthropy. And you probably don't work for a poverty nonprofit.

So how can you support the end of bootstrapping and the rise of Community with a big "C"?

On the smallest level, each one of us can start by rewriting our own narratives about achievement and attainment. We can each question how we both publicly and privately tend to attribute our successes to our own abilities alone, how we unwittingly absorb the just-world bias into our daily lives.

How do we change this? For starters, we can try to revise our framing of our own existences. We can acknowledge to our friends and colleagues both our dependence and interdependence, our deficiencies and our economic struggles but also unspoken privileges. In other words, we can be transparent about our own messiness and how we have gotten where we are by leaning on others or social systems.

One way we can do the latter is to give public, verbal credit when we can to those who have helped us along the way, including the people who may have taken care of our children. We can ask ourselves: *Who assisted in your success?* (I learned this technique from a podcast host I met in my travels.) On a small level, these expressions of thanks puncture the triumphal and false story line of individual success. When I do this exercise, I think of what I owe my grandparents, who ran a small shoe store in the Bronx, which meant I was literally playing—probably this book's title was not a coincidence after all—with boots and shoehorns, polish, and shoe brushes on the floor of their apartment every weekend. I also recall the writing teachers who helped me early on in my teen years, including the author Frank McCourt, my English teacher at Stuyvesant High School, and his kindly comments on my papers. You will likely have your own list, all more evidence of the art of interdependence, as any singular achievement has thousands of other people behind it. This practice may sound a little woo-woo, like burning sage in an apartment to cleanse it of bad energy, but it's oddly reassuring.

There is also another kind of personal liberation from the con of pure independence: expressing our distaste for the "bootstraps" story line to others. We can follow the example of Michelle Hughes at the National Young Farmers Coalition, who writes: "Questioning 'bootstrap' and other destructive myths is the first step. From here on out I am committing to writing a new narrative that respects all of our struggles and truly embodies what I believe in." The bootstrapping ideology, Hughes continues, particularly for a Black woman and a farmers' advocate like her, is "triggering and deeply damaging."

In addition, if we rely on systems that should be providing enfranchisement, from welfare to health insurance payments to voter registration, but often aren't, we need a greater awareness of the administrative burden that was often laced into these systems, making them badly made from the start and sometimes on purpose. Once we recognize the mechanism that turns applying for most forms of governmental assistance into an ordeal, we can forgive ourselves for being exhausted by it. We can also push to make the pandemic period's efficiencies added to these systems, including Medicaid and SNAP, permanent. As I have mentioned, enrollment in Medicaid has been up by 20 percent since the pandemic began, partially because people hadn't been cycled on and off the aid program as the constant process that called for people who are enrolled to recertify had been paused. Pausing or simply slowing recertification and getting rid of in-person interviews for SNAP from now on would seriously streamline these procedures, giving people who require the aid access to it.

We can also support activists among those who are financially stressed, who instead of bowing to impediments have put themselves on the line. These include those fighting for housing justice through rent strikes and also through imaginative and theatrical efforts, like KC Tenants in Kansas City, an organization led by self-described poor and working-class individuals who protest tenants (including sometimes themselves) being tossed out of their homes. That group's activities also include the piquant "Slumlord Saturdays," a social media campaign that publicized delinquent landlords who were pressing for evictions. And their efforts have often been effective: in response to this pressure, the city council in Kansas City passed a Tenant Bill of Rights.

In the face of the proliferation of this sort of mutualism and activism as well as a series of political setbacks, I have started, as a game, to think of how this might play out in the future. I fast-forward in my mind's eye to a time far from now. Like a sci-fi television showrunner, I imagine what it will be like in 2040 or 2050. The clichéd view is that it will be bleaker, or maybe absurdly technological—a sea of robot children or rows of AI drones. Yet I tend to imagine that this to-be era will be a little bit rosier than all that. In 2050, I wager, the dystopian social safety net might be like a relic from the Dark Ages, with far fewer basic health needs answered by a grim crowd-funding platform or whatever we will call such vehicles in the future. I also imagine far less commuting to bullshit jobs or exhausting work-related travel. There might be more strolls around local ponds, more charades with your kids, and more focused work, undiluted by advantage-seeking lunches and jostling happy hours with detached, beery colleagues.

What must happen to arrive at this state, though, goes beyond the beloved broad-based movements and familiar narratives. In terms of the latter, there are other stories we can choose from. We might embrace the notion of "secular faith" that one scholar has argued for, engaging politically and personally within our community and the larger community that is our country. (Translation: Let's put our petty existences at the service of public projects!) We can reach for the warmth and amplitude created when we are linked to and engaged with one another and the outside world. We could use a few other affectionate names to describe this kind of commitment to something outside of ourselves, phrases like communal luxury or public happiness. Whatever we call it, it will offer us the richness of a life that is better led and of a more capacious political imagination.

We might commit ourselves, in fact, to such electric social or political feeling. If we do rush toward these emotions, we might, at the very least, be able to encounter "new main characters, better plots, and at least the possibility of some happy endings," as one activist writer puts it, the happiness being likelier to arrive "when we are able to take joy in the flourishing of others." At the very least, we may discover in this new passion, becoming radiant with expressiveness.

In order to get to this more radiantly centered place, though, we each need to work to snuff out the self-made myth within ourselves and to stop flagellating ourselves when we don't achieve autonomous glory. We must accept our dependence, celebrating ourselves not only when we stand on our own two feet, as the saying goes, but also when we ask for help or rely on a federal social program for assistance or simply achieve a goal as a group rather than on our own.

We might instead see ourselves as taking part in a broader experience, recovering with others rather than alone—finally, in truth, becoming part of a society. Needing one another is our strength, not our weakness. Singular triumphs never existed in the first place.

Epilogue: My Interdependence Day

IN THE SPRING of 2020, my internalized taskmaster took over. At first, I baked loaves of whole wheat bread in a Dutch oven almost every day, instead of buying them, to avoid stores, as if becoming militant at the arts of idling might help me nibble at the edges of calamity. After my daughter's schooling shifted to remote learning, I tried to find order in this new regime by constructing color-coded schedules for her, my own form of self-improvement for both her and me. After cooking, educating, and working, I'd run miles outdoors in the chilly weather. I'd listen to the very basic New York birds' songs and memorize the barks of the newly budding trees. And then I attended to my daughter, who wheedled for a puppy: I wrote pieces of this book while my daughter was shouting these requests from the other room. To distract her, I introduced her to neighbors' dogs to get some relief: I was then able to type a few more paragraphs at a time.

Our family was lucky as the months went by, in that we didn't sicken and our jobs didn't fall away from us: we were privileged enough, also, not to have to work from an office with hot desks or a grocery store, as some of the people who wrote for me about their pandemic experiences had to.

I wasn't at a formal work site but that didn't mean I wasn't often, in some way or other, working. In fact, I soon labored, in one way or another, almost all the time. As the pandemic progressed, a casual run became an effortful six-mile race, mostly uphill. The loaves of bread, at first simple gummy mounds, had to become more. I soon began quizzing myself on conifer names and those of obscure flowers and small animals. It seemed I couldn't help but lay out any information on a grid of achievement, trying to hack self-betterment.

As for the dog, when we finally caved and obtained a puppy for my daughter, my inner drill sergeant took over again, this time regarding the canine. I was the one mixing the corn syrup in our pup's bowl as the fragile beast had hypoglycemia. Then, one hand on the laptop and the other feeding the dog pinkie by pinkie. I even had to prove myself worthy *to the dog*.

This is my own version of the distortions that I write about in these pages. I have since I was young attempted to achieve my own version of the American Dream. I'd absorbed my own variation of bootstrapping, it seemed, with the mother's milk of my native country.

I sometimes felt like the drawing in a bestselling illustrated memoir: the protagonist shows her ripped biceps while working out, and the caption reads: "The next step in my self-improvement program . . . is to embrace my interdependence." (She and I were advantaged enough to take our own achievement that seriously, in the first place.)

Turning what should be casual hobbies into severe pastimes has long been one of the ways I have been guilty of self-punishing, a version of the insistence that we succeed on our own.

And then, I started to ease off.

It began when my daughter, once an exuberant child, was suddenly sad from being separated from her teachers and friends: I hoped that even gesturing at community connection might be one answer for her as well. She and I started to make watercolor drawings of our neighbors' dogs and dropped them off at their houses along with painted small rocks that my daughter crafted, which read "Stay Well" and "Don't Cough," all in silver paint in her looping script. During the first pandemic summer, we took long walks to drop off canned chickpeas and apples at the mutual aid food pantry. Later on, during her time of hybrid schooling, my daughter spent the school day at her friend's house, and her friend spent another school day at our house, a practice that she understood, in her own way, to be interdependent. It turned out that my elderly mother was having her own take on the new interconnection. Locked in her home in western Massachusetts, she awaited the mutual aid worker with a weekly drop-off of birdseed so that she could watch birds from her window, that rare pandemic pleasure.

I attended a Zoom Passover, one of the many rituals taking place virtually, replete with awkward pauses, flickering screens, and choppy pixels; with colleagues or friends or audience members' heads bobbing on-screen, unclear who was the speaker, who were the listeners, and who was the chorus, like *Hollywood Squares* by way of Kafka. I reveled in communicating with the people in this book as well, whether over an old landline or online in pixelated encounters. I spoke to those in my extended network, some who were mostly alone. These exchanges were throwbacks. Even as I was interacting like this, these episodes seemed to have a grainy, vintage hue, as if they were already memories. The tragedy was mutual also, as when I attended two Zoom shivas in honor of men who had died of COVID.

As I told my daughter then, each one of us may exhibit these three states—interdependence, dependence, and independence—at different points over the course of a month, a day, or an hour. As an example: one day, our family helped a friend of ours cover her rent during a particularly hard week. Another day, a friend cried to us about her pain. A third day, we accepted a gift from a family member.

Within a single week, with my friends, I exulted in small moments of connection. One of my colleagues made me a mixtape; a neighbor dropped off tarragon. A Facebook friend in Sarajevo early in the pandemic posted that "whenever we make a nice meal, we share it" with neighbors, including the long-saved can of foie gras she'd gotten for her birthday. I tried a similar model. When I encountered friends "in the wild," we sat down on the ground of a park with each other, like kindergartners, while observing proper social distancing. When I had my birthday during quarantine, my daughter and I and other friends walked, each of us at a safe distance from one another, facing the wind and the sun. We were mutually linked in these moments, like threads in a fleshly safety net—providing emotional support while trusting ourselves to stay physically apart.

When listening to catbirds' outrageous songs, I was suddenly back within the original purpose of birdwatching, just being in the natural world, rather than trying to get a proverbial letter grade A in avian life. Paradoxically, I was partially able to let go of the habits of mind of individualism *because* of the privileges I had been born with, as a white woman of a middle-class background, and the advantages I had acquired, like a white-collar job with flexible hours. These ultimately allowed me the mental freedom to reconceive of what I personally found valuable during that time. I know that not everyone has this kind of space by which to recalibrate a relationship with the bootstrapping society that pinions so many of us.

But once I did reconsider my own version of individualism, I experienced my connections to the interlocking communities around me all the more strongly. They existed like a fantasy body in my mind and sometimes like real bodies in space, which I sought to stay linked with, and to serve. Serving a community, no matter how loosely knit, started to feel like serving a larger cause. I thought of a term used by feminist philosophers such as Marilyn Friedman, "relational autonomy"—I was seeking to obtain a state of independence where I was at the same time concatenated with others.

I read about *dependence* as a word once having meant "hanging down" or "hanging from" another. That etymology helped me, as I became more interdependent during the pandemic and, yes, more dependent on others. It was one of many paradoxes that togetherness flowered amid germ-fearing isolation.

In Brooklyn, during a socially distanced visit to the Prospect Park Zoo, I sat and watched the primates and saw in them something different than I had before: a version of a mutual aid group. These were gray, crimp-haired Hamadryas baboons on one of the first days in 2020 that their zoo home was open again, playing with one another in the water and grooming one another for hours. They were traveling nearly in unison around their habitat. I kept expecting them to fight each other, but they never did. We onlookers stared at these communicative, caring baboons with their retro-glam-rock gunmetal-hued fur. The timbre of these baboon lives was what interdependent humans might experience for themselves (ideally, uncaged).

Solidarity spreads infinitely outward, creating beneficial eddies for all. In contrast, the bootstrap lie is finite. Its bounties do not "trickle down" or spread outward as those who believe the self-made narrative profess that they do.

Toppling the self-made myth starts with each of us telling divergent tales about America, presenting our own stories within that larger story differently. These in turn expand outward into both what we do and what we expect from our government. This moment has offered us a new narrative of collective support, which carries possibility. The social and economic benefits can only grow, with each one of us beginning to see our lives and our country through this shared vista. We may one day look back on these years as the time when we awakened to a new way of being—when looking out for ourselves and looking out for one another became one and the same thing. It's up to us to continue what we began out of necessity. As I tell my daughter, "We take care of each other." She is starting to believe me.

Acknowledgments

MOST BOOKS ARE testaments to both the effort and vision of individual authors and the collectives of which they are a part. This book is no exception: it, too, is the beneficiary of the very interdependence celebrated in its pages. Fortunately for me, my life is interwoven with—and I am dependent on—a vast array of gifted writers and editors.

Thanks first off to my consistently lovely editor, Gabriella Doob, and appreciation to my beloved agent, Jill Grinberg, and her colleague Denise Page. Gratitude in advance to the publicity team at Ecco, headed by Miriam Parker, who are reliably excellent. And thanks to Denise Oswald for seeing the promise of this book from the get-go.

I'd also like to thank my friend Maia Szalavitz, who read many drafts and was a constant force of editorial care and reassurance. I'd like to credit two Anns, Ann Neumann and Ann Peters. They both bravely read early drafts and served as the most thoughtful interlocutors. Ditto for the wonderful Kim Cutter, who prompted me with great queries, including while we went on long pandemic walks, all while the book was still incomplete. As the book progressed, the inimitable Genevieve Field offered beautiful line edits, as did Julie

Lasky. In addition, Julie assisted in the concept of the cover, as did ace designers Scott Menchin, Yvetta Fedorova, Penny Blatt, and my dearest artist friend Rachel Urkowitz. John Webb and Kyla Jones provided inspired and adept copyedits and research help, respectively.

In addition to the most active readers, I'd also like to thank the many friends I've discussed this book with over the years who offered a variety of reporting, historical, and even chapter title suggestions, including the cherished Jared Hohlt, Elena Krumova, Anne Kornhauser, and Hamilton Nolan. Additional advice was offered by prized pals Laura Secor, Sarah Safer, Lauren Sandler, and Kathy Stewart. My mother, Barbara Koenig Quart, has been a stalwart as well.

Appreciation also goes to my incredible colleagues at the Economic Hardship Reporting Project, most crucially David Wallis but also to Duy Linh Tu and Jane Englebardt, among others. A hat tip also to the Logan Nonfiction Foundation, which provided a fellowship, where I was given a week in the country, free from work and family, to complete a round of final edits. I'd be remiss also if I didn't name-check some of the editors and granters who supported pieces of reporting that went into this and the idea of the book overall, from Maddie Oatman at *Mother Jones* to the late, excellent Chris Shea at the *Washington Post* to Sarah Gustavus Lim and Alexis Krieg. I must also praise Barbara Ehrenreich, whose legacy and writing have fundamentally shaped not only my reporting and mindset but also my professional existence.

Let me acknowledge as well how much I owe to the generosity and honesty of my sources for the creation of this title. I think of Cissy White here among many others. I am grateful that you shared your stories with me.

Finally, gratitude to my husband, the great writer Peter Maass, and my child, the great daughter Cleo Quart Maass. You are the mutual aid society that is our home. I love you immensely.

Notes

Preface

xii the antique sociological concept of "organic solidarity": The nineteenth- and early-twentieth-century French social scientist Émile Durkheim argued there were two kinds of solidarity. The first was mechanical—the cohesiveness of smaller and more homogenous and tribal societies. The second was organic, where more complex and diverse societies are brought together through a more complicated division of labor.

Chapter 1: The Backstory

5 As the broadsheet *Working Man's Advocate* put it: A newspaper published a squib, "KNOW YE, that I, NIMROD MURPHREE, of the city of Nashville, and state of Tennessee, have discovered perpetual motion. N. MURPHREE," on September 30, 1834. This article was in turn parodied in a piece in *The Working Man's Advocate* on October 4, 1834.

6 The absurdist use of the idiom: *The Dial: A Monthly Magazine for Literature, Philosophy and Religion*, M. D, Conway, editor, 1860, volume 1, https://books.google.com/books?id=yUIUAAAAYAAJ&pg=PA709&lpg=PA709&dq=Sir+William+Hamilton+bootstraps&source=bl&ots=fl0y7CIK6v&sig=ACfU3U1Crtqhc3nbUlGa2HMLgtXeY86_9g&hl=en&sa=X&ved=2ahUKEwiH0Za3mvb1AhWmVt8KHVRDAFwQ6AF6BAgiEAM#v=onepage&q=Sir%20William%20Hamilton%20bootstraps&f=false.

6 "The attempt of the mind to analyze itself": Jan Freeman, *Boston Globe*, January 25, 2009, http://archive.boston.com/bostonglobe/ideas/articles/2009/01/25/the_unkindliest_cut/?page=2.

6 "self-made man," as popularized by the Kentucky politician Henry Clay in 1832: "Clay famously used the phrase 'self-made men' in the US Senate while defending the American System, which advocated federal support for manufacturing and infrastructure. Clay praised Kentucky manufacturers as 'enterprising and self-made men' who deserved the nation's favours because they had 'acquired whatever wealth they possess by patient and diligent labor." Pamela Walker Laird, "How Business Historians Can Save the World—from the Fallacy of Self-Made Success," *Business History*, November 9, 2017, https://www.tandfonline.com/doi/abs/10.1080/00076791.2016.1251904?journalCode=fbsh20.

7 including "welfare queens": Reagan's anti-entitlement conservatism also, ironically, attracted many of the same World War II veterans who benefited from the GI Bill. These vets were among "the core [Reagan] constituency opposing taxpayer funding of social programs, with the result that only meager benefits await those returning from today's wars," as the journalist Edward Humes wrote. As one GI Bill recipient put it to the writer Studs Terkel: "It was bad in a way. A lotta people just sat, they didn't even look for jobs. Just like welfare." In addition the GI Bill provided many working-class young men with educations and helped them buy homes, yet it didn't offer the same help for Black veterans.

7 "a cancer eating at our vitals": "The Original 'Welfare Queen,'" NPR, *Code Switch*, June 5, 2019.

7 Tim Boyd, the now-former mayor: "A Former Texas Mayor Said Residents Should Fend for Themselves," Christine Hauser, *New York Times*, February 18, 2021, https://www.nytimes.com/2021/02/18/us/tim-boyd-mayor-colorado-city-texas.html.

8 in group efforts like barn raisings: In the 1914 account in the book *Rural Manhood*, I read how in eastern Pennsylvania when one family required a new barn for themselves and their 150 cattle, they hired a single carpenter and that carpenter created a barn-raising party where the men arrived at six in the morning from a radius of four miles away and brought their picks and shovels, and teams of horses and heavy wagons, ready to dig an enormous trench. They would lay down twenty-seven thousand feet of lumber for framing and five thousand feet of flooring. The barn was ultimately sixty-five feet high, one of thousands of raisings that occurred in that time, and part of a collectivist impulse often within religious sects in the Northeast.

9   "color a stigma": From a taped interview with Martin Luther King Jr. with NBC News Sander Vanocour at Atlanta's Ebenezer Baptist Church, May 8, 1967, less than a year before his assassination. https://www.prime timer.com/watch/the-rarely-seen-1967-nbc-news-interview-of-dr-martin -luther-king-jr.

10 tax rate of 91 percent: Bernie Sanders announced this fact in November 2015 in Des Moines, Iowa. According to the Tax Foundation's federal income tax rates history, during the eight years of the Eisenhower presidency, from 1953 to 1961, the top rate was 91 percent.

Chapter 2: Where's Walden?

15 As political philosopher Nancy Fraser: In a 1994 essay, they write that the bias against dependence in the nineteenth century could be considered racist. Those who were "dependent" were forced to be so either legally or due to social norms—they were slaves, colonial natives, and women who worked in the home. Nancy Fraser and Linda Gordon, "'Dependency' Demystified: Inscriptions of Power in a Keyword of the Welfare State," *Social Politics: International Studies in Gender, State & Society* 1, no. 1 (Spring 1994): 4–31, https://doi.org/10.1093/sp/1.1.4.

17 In the summer of 1836, for instance: Descriptions of some of the Transcendental Club meetings derive from Samuel A. Schreiner Jr.'s book *The Concord Quartet: Alcott, Emerson, Hawthorne, Thoreau, and the Friendship That Freed the American Mind* (Hoboken, NJ: Wiley, 2006).

17 "suddenly well and strong . . . [Thoreau is] as full of buds of promise as a young apple tree": This letter from Emerson is reprinted in Jeffrey S. Cramer, *Solid Seasons: The Friendship of Henry David Thoreau and Ralph Waldo Emerson* (Berkeley, CA: Counterpoint Press, 2019).

18 He inherited $11,600 in 1834: From Albert J. von Frank's *An Emerson Chronology* (Boston: G. K. Hall, 1994), 91.

18 Emerson's second wife, Lidian: This detail is from the fine and Emerson-loving biography by Richard D. Richardson, *Emerson: The Mind on Fire* (Oakland, CA: University of California Press, 1995).

19 "imperial self": Quentin Anderson, *The Imperial Self: An Essay in American Literary and Cultural History* (New York: Knopf, 1971).

19 As scholar James Read writes in his essay: "The Limits of Self-Reliance: Emerson, Slavery, and Abolition" was presented by James H. Read to the annual meeting of the American Political Science Association, Toronto, September 3–6, 2009, detailing how Emerson "saw antislavery activism as a distraction from his own proper work of freeing 'imprisoned spirits, imprisoned thoughts, far back in the brain of man.'"

19 the critic Leo Marx writing scathingly: Irving Howe, reply by Leo Marx: "Emerson and Socialism: An Exchange," *New York Review of Books*, May 28, 1987, https://www.nybooks.com/articles/1987/05/28/emerson -and-socialism-an-exchange/.

20 "the captain of a huckleberry party": From the book by Scott A. Sandage, *Born Losers: A History of Failure in America* (Cambridge, MA: Harvard University Press, 2006), and my interview with Sandage.

22 Mrs. Rowson's Academy for Young Ladies in Boston: Clare Hunter, *Threads of Life: A History of the World Through the Eye of a Needle* (London: Hodder & Stoughton, 2019), 190–91.

22 The students at Susanna Rowson's famed school: The school's creator, Susanna Rowson, also an actor and an author, offered the girls design sources from literature and history so they might improve their minds while doing the embroidery, considered appropriate women's work.

23 labored in anonymity: At Rowson's Academy, for instance, in 1806, Narcissa Sewall, then fifteen years old, created needlework pictures in fine silk thread. She sat, head bent, surrounded by other girls and young women. One needlework picture depicted the poem "The Friar of Orders Gray"— when she wasn't stitching, she was memorizing and reciting the poem. Sewall wasn't the only teenager making fine craft based on literature. An effort by one of Sewall's contemporaries was based on a painting of a theatrical production of *Antony and Cleopatra*. Now Sewall's work is in the Metropolitan Museum of Art: https://www.metmuseum.org/art/collection /search/746465.

Chapter 3: Little House of Propaganda

26 "poisonous cultural worship:" Alex Acks, "The Weird Libertarian Trojan House That Is Little House," *Book Riot*, November 26, 2018, https:// bookriot.com/the-weird-libertarian-trojan-horse-that-is-the-little-house -books/.

27 In 1928, Hoover used the phrase "rugged individualism": He used it in a 1928 campaign speech. The phrase "American Individualism" can be found in his book of the same name. From the book: "The American pioneer is the epic expression of that individualism, and the pioneer spirit is the response to the challenge of opportunity, to the challenge of nature, to the challenge of life, to the call of the frontier. That spirit need never die for lack of something for it to achieve," https://www.hoover.org/research/future -american-individualism#:~:text=The%20American%20pioneer%20 is%20the,something%20for%20it%20to%20achieve.

28 "We have a dictator," Lane wrote of Roosevelt in her journal: Judith Thurman: "Wilder Women: The Mother and Daughter Behind the Little

House Stories," *New Yorker*, August 3, 2009, https://www.newyorker
.com/magazine/2009/08/10/wilder-women.

28 It supported 6,600 writers: Douglas Brinkley, "Unmasking Writers of
the W.P.A.," *New York Times*, August 2, 2003, https://www.nytimes
.com/2003/08/02/books/unmasking-writers-of-the-wpa.html. In addition
to helping journalists survive, FWP work often came in original formats:
for example, federally funded reporters produced an American Guide
Series for city and state travel.

30 and those like them: Oklahoma Historical Society, "The Final Period,
1867–1892," https://www.okhistory.org/research/airemoval. The Kansa-
pedia or the Kansas Historical Society site writes of a tribal population
killed and starved by US treaties, that in that time "declined from several
thousand to 1,500 by 1800, to 553 by 1872, and to 194 within 16 years
of the 1873 move to Indian Territory (present-day Oklahoma)," https://
www.kshs.org/kansapedia/kaws-or-kanzas-kansas/17371. Accessed February 2022.

30 "Was the American frontier 'conquered' by single scouts": Carol Tavris,
The Mismeasure of Woman (New York: Touchstone, 1993).

31 "big-veined and almost black": Agnes Smedley, *Daughter of Earth*, reprint
(New York: The Feminist Press at CUNY, 1993).

32 different notion of independence than the one propagated by *Little House*:
My daughter had never shown any interest in the romantic immigrant
children's literature that was *Little House*'s correlative, all books I consumed with fervor when I was young. They were centered around cheerful
hardworking kids on the Jewish Lower East Side of yesteryear, including
titles like *How They Grew* by Margaret Sidney.

32 Sunaura Taylor: Joshua Rothman, "Are Disability Rights and Animal Rights
Connected?," *New Yorker*, June 5, 2017, https://www.newyorker.com/cul
ture/persons-of-interest/are-disability-rights-and-animal-rights-connected.

32 the Homestead Act of 1862: Kerri Leigh Merritt, "Race, Reconstruction,
and Reparations," *Black Perspectives*, February 9, 2016, https://www.aaihs
.org/race-reconstruction/.

33 an origin point of Black landlessness: The number of adult descendants
of the original Homestead Act recipients in 2000 has been estimated at
around forty-six million Americans.

34 a "massive transfer of wealth": In *Freedom from the Market*, Konczal writes
that the word *freedom* should be used to describe "free land," or "free
time," rather than "freedom" from, where we are supposedly entirely independent of our society. Mike Konczal: *Freedom from the Market: America's
Fight to Liberate Itself from the Grip of the Invisible Hand* (New York: The
New Press, 2021).

34 he was not such a great farmer: This is described in Caroline Fraser, *Prairie Fires: The American Dreams of Laura Ingalls Wilder* (New York: Henry Holt, 2017). Pa Ingalls, Fraser writes, also had a "moral ambiguity missing from the portrait his daughter would one day so lovingly polish."

34 Vernon Parrington, called "the great barbeque": Colin Gordon, "The Great Barbecue Revisited," *Dissent*, July 4, 2013, https://www.dissentmagazine.org/blog/the-great-barbecue-revisited.

35 NBC's highest-rated scripted program: Joel Swerdlow, "TV's Ringing Hit in the Heartland," *Washington Post*, May 25, 1980, https://www.washingtonpost.com/archive/lifestyle/1980/05/25/tvs-ringing-hit-in-the-heartland/29d75461-217e-4596-b9a3-901926c0a42d/.

35 cover of *TV Guide* twenty-two times: "Landon appeared on the cover of TV Guide 22 times, second only to Lucille Ball," "Michael Landon," Television Academy, https://www.emmys.com/bios/michael-landon.

35 Reagan and Michael Landon had a similar suntanned affability: Reagan and Landon represented differing yet overlapping values my leftist parents found both artificial and reprehensible.

Chapter 4: The Horatio Alger Lie

40 his fourth young adult book: Horatio Alger, Jr., *Ragged Dick: Or, Street Life in New York with the Boot Blacks* (Boston: A. K. Loring, May 1868).

41 they help him make it off the streets: In *Ragged Dick*, wealthy men give him a new suit to replace his tattered clothes. Dick seeks out tutoring and opens a bank account. In a strange twist, he winds up with a job at a business firm. No longer a vagabond, he changes his name from the low-living, informally boyish "Dick" to the grander "Richard." This was what the Alger specialist Jeffrey Louis Decker calls "market pluck."

41 Alger's boys "escape precarious financial circumstances": Alger scholar Nackenoff told me in an interview that there "is really no role for government beyond the police in [Alger's] stories." Everything happens without government intervention, which while today might be called libertarian is also a distinctly pre–New Deal worldview. Some of Nackenoff's insights derived from interviews with her, but also from her book *The Fictional Republic: Horatio Alger and American Political Discourse* (New York: Oxford University Press, 1994).

41 not the one found in the well-known Ralph Gardner biography: Ralph D. Gardner, *Horatio Alger: Or, the American Hero Era* (London: Arco Publishing Company, 1978).

42 As James Martel, a professor at San Francisco State University: I refer to both his paper "Horatio Alger and the Closeting of the Self-Made Man" and an interview I conducted with the professor of political theory.

43 He spent his free time observing these homeless kids: Information presented by the Institute for Children, Poverty, and Homelessness, http://nyc homelesshistory.org/era/nineteenth/.

44 Republican-leaning people say that "hard work" is the explanation behind wealth: "Most Americans Point to Circumstances, Not Work Ethic, for Why People Are Rich or Poor," Pew Research Center, March 2, 2020, https://www.pewresearch.org/politics/2020/03/02/most-americans-point -to-circumstances-not-work-ethic-as-reasons-people-are-rich-or-poor/.

44 In 2015, while Wanek was at the helm: From a 2015 OSHA news release, accessed on March 4, 2022, https://www.osha.gov/news/newsreleases /region5/02022015.

46 entitled *Mean Girl,* writes: Lisa Duggan, *Mean Girl: Ayn Rand and the Culture of Greed* (Oakland, CA: University of California Press, 2019).

46 Rand inspired Whole Foods Market's CEO John Mackey: John Mackey, "The Whole Foods Alternative to ObamaCare," *Wall Street Journal,* August 11, 2009, https://www.wsj.com/articles/SB100014240529702042 51404574342170072865070.

47 and was contributor to her book *Capitalism: The Unknown Ideal*: The subtitle of this book cracks me up, as capitalism can seem to be the only ideal Americans universally know. Ayn Rand, *Capitalism: The Unknown Ideal,* first edition (New York: New American Library, 1966).

48 Rand's biographer Anne Conover Heller depicts: Anne C. Heller, *Ayn Rand and the World She Made* (New York: Anchor Books, 2009).

49 The setting of *The Fountainhead*: *The Fountainhead,* King Vidor, director, Warner Brothers (1949).

49 The usually reliable director King Vidor: Raymond Durgnat and Scott Simmon, *King Vidor, American* (Oakland, CA: University of California Press, 1988), 263. The writers put it like this: the film borrows "*film noir*'s angles and darkness, its paranoia." Film noir here is considered a befitting form for Rand's over-the-top individualism because noir focuses on the one person, trapped by circumstance (and by shadows). As they write, "Every man walks alone down dark, mean streets."

49 In its time, the film, deservedly, bombed: Perhaps I should have tried to watch the film earlier in the day, as my eyes fluttered shut during the film's chain of absurdly didactic and stilted scenes. The high point was when Neal's character, Dominique, cheekbones as sharp as the film's architectural monstrosities, slaps Cooper's Roark with a riding crop.

50 a "privatized state": Political scientist Chiara Cordelli explains the privatized state that we dwell in today as distinct from that of the last century, with its bureaucrats, ministers, public officials, and civil servants—today's America emphasizes corporate fixes and institutions at every turn. Chiara

Cordelli, *The Privatized State* (Princeton, NJ: Princeton University Press, 2020).

Chapter 5: Rich Fictions

56 "Well, one day I'd like to go to the moon": Paul Orfalea says this in the 2006 documentary *The One Percent*, directed by Jamie Johnson.

57 earlier in the last decade was 354 to 1: Sorapop Kiatpongsan and Michael I. Norton, "How Much (More) Should CEOs Make? A Universal Desire for More Equal Pay," *Perspectives on Psychological Science* 9, no. 6 (2014), https://www.hbs.edu/ris/Publication%2520Files/kiatpongsan%2520norton%25202014_f02b004a-c2de-4358-9811-ea273d37 2af7.pdf.

57 Thomas Piketty estimated that roughly 60 percent: This wealth comes in the form of savings, houses, and investment, that "people with inherited wealth need save only a portion of their income from capital to see that capital grow more quickly than the economy as a whole." Thomas Piketty, *Capital in the Twenty-First Century* (Cambridge, MA: Harvard University Press, 2014).

57 At the same time, Americans think their odds of success: Sweden, the home of copious social welfare, nonetheless has pessimistic citizens. The Swedes believe their chance for mobility from the bottom quintile to the top to be roughly 9 percent, according to research by the economists Alberto Alesina, Stefanie Stantcheva, and Edoardo Teso, when the likelihood of success was actually better: 11 percent.

58 European counterparts: In contrast, Americans were much more optimistic about their potential mobility. They also had far less reason for bullishness. In their 2018 study in *American Economic Review*, "Intergenerational Mobility and Preferences for Redistribution," the other pessimistic, mobility-doubting countries in the study included Italy and Sweden, and the authors found "strong political polarization. Left-wing respondents are more pessimistic about mobility: their preferences for redistribution are correlated with their mobility perceptions."

58 "To say that Steve Jobs didn't build Apple": Romney's line about Papa John's was clearly part of the Romney campaign's response to then–President Barack Obama's speech the previous week in Virginia, where he said, "If you've got a business—you didn't build that," implying how interdependent entrepreneurs and workers are, but it was also just as clearly part of the rich fiction of self-propulsion.

58 as Robert Reich suggested: Robert B. Reich, "Entrepreneurship Reconsidered: The Team as Hero," *Harvard Business Review*, May 1987, https://hbr.org/1987/05/entrepreneurship-reconsidered-the-team-as-hero.

58   the annual *Forbes* billionaires list in April 2021: "Billionaires Club Grew by Nearly a Third, to 2,755, During Pandemic," Hannah Denham, *Washington Post*, April 6, 2021, https://www.washingtonpost.com/business/2021/04/06/billionaire-wealth-forbes-pandemic.

59   psychiatrists now call a "moral injury": "What Is Moral Injury," The Moral Injury Project, Syracuse University, accessed September 8, 2021, https://moralinjuryproject.syr.edu/about-moral-injury/.

59   Gregg Gonsalves tweeted in 2020: In addition, Gonsalves tweeted, "the small and sundry Generation X were told all of our lives, starting with Reagan's presidency, that the 'government is the problem.'"

59   through everything from political ads and business cable shows to think tanks: If we consider think tanks, we might look at Charles Koch and his brother, the late David Koch. The Koch brothers funded the Cato Institute and the Institute for Humane Studies, which describes itself as a "nonprofit educational organization that engages with students and professors around the country to encourage the study and advancement of freedom," injecting many millions a year into several hundred US universities to encourage libertarianism. Possessed by anti-government mania, Cato and others argue that we are supposed to take care of ourselves, and we should be neither assisted nor empathized with in our downward descents.

60   According to Subsidy Tracker, Tesla received nearly $2.5 billion: Or, as Jerry Hirsch's *Los Angeles Times* headline reads, for all of Musk's companies at that time, according to their data: "Elon Musk's Growing Empire Is Fueled by $4.9 Billion in Government Subsidies" (May 30, 2015).

61   individual effort in success: In a similar vein, Ivanka Trump's "motivational quotes" during her father's presidential tenure included "If you are content, that's probably not good enough," faulting women for not being able to surmount obstacles, as if they were ne'er-do-wells creating roadblocks for themselves, the glass ceiling all in their heads.

61   feminist Sheryl Sandberg puts it in her 2013 bestseller *Lean In*: Sheryl Sandberg, *Lean In: Women, Work, and the Will to Lead* (New York: Knopf, 2013).

61   a cute neologism coined by the founder of the online women's retailer Nasty Gal, Sophia Amoruso: Sophia Amoruso, *#GIRLBOSS* (New York: Portfolio, 2014). She poses on the cover of one edition in a tight black dress with a plunging neckline and her hair so perfectly straight it might well be a Lulu-Brooks-goes-Instagram wig.

62   from scratch themselves: A villainous version of the "girlboss" was Elizabeth Holmes, the mastermind behind the scam blood-testing company Theranos.

62   as Andi Zeisler writes: Andi Zeisler, *We Were Feminists Once: From Riot*

Grrrl to CoverGirl®, the Buying and Selling of a Political Movement (New York: PublicAffairs, 2016).

63 As the scholar Tressie McMillan Cottom puts it: Tressie McMillan Cottom, "Trickle-Down Feminism, Revisited: 'Having It All' Is Not a Feminist Theory of Change," *Dissent*, April 21, 2016, https://www.dissent magazine.org/blog/anne-marie-slaughter-trickle-down-feminism-un finished-business-review.

64 Gelman's plush velvet-and-brass bootstrapping form of feminism: Amanda Hess, "The Wing Is a Women's Utopia. Unless You Work There," *New York Times Magazine*, March 17, 2020, https://www.nytimes.com/2020/03/17/magazine/the-wing.html.

65 So Lerner began a study: Melvin J. Lerner and Carolyn H. Simmons, "Observer's Reaction to the 'Innocent Victim': Compassion or Rejection?," *Journal of Personality and Social Psychology* 4, no. 2 (1966): 203–10, http://web.mit.edu/curhan/www/docs/Articles/biases/4_J_Personality_Social_Psychology_203_(Lerner).pdf.

66 characterized the woman negatively: Within the experiment, participants were asked to vote for an action that should be directed at the victim: negative reinforcement, positive reinforcement, or control. At some point, the women were also asked to describe their impression of the victim's personality and asked how attractive she appeared and how much they identified with—or saw themselves in—her. These written comments from the college students led to the most "interesting data," wrote the researchers.

66 that other people get what they deserve: We tend to believe that if we try hard enough for the best education, we will attain our hopes and dreams, and we are told to ignore the disadvantages around race and class woven into our public education system that make achieving our ambitions too rarely the outcome. The just-world hypothesis flourishes because we tend to blur out the disadvantages that our social differences have created for us, and if a person is not solvent or successful on paper, they are somehow inherently weaker. This leads to disturbing related beliefs and attitudes. For instance, those who believe in a "just world" also tend to disapprove of affirmative action.

67 The just-world theory: "Americans Overestimate Social Mobility in Their Country," *Economist*, February 14, 2018, https://www.economist.com /graphic-detail/2018/02/14/americans-overestimate-social-mobility-in -their-country.

69 "Somebody will say, 'We have a problem'": This remark was in archival footage in the fine 2020 Showtime documentary series *The Reagans*.

Chapter 6: The Self-Made Voter

73 After hearing of a spate of farmer suicides: Katie Wedell, Lucille Sherman, and Sky Chadde, "Midwest Farmers Face a Crisis. Hundreds Are Dying by Suicide," *USA Today News*, March 9, 2020, https://www.usatoday.com/in-depth/news/investigations/2020/03/09/climate-tariffs-debt-and-isolation-drive-some-farmers-suicide/4955865002/.

73 Roecker estimated that on average farmers in his town were each losing $30,000 a month: According to the Department of Agriculture, Wisconsin lost 10 percent of its dairy farms in 2019, a total of 819. Joseph Zeballos-Roig, "Wisconsin Lost 10% of Its Dairy Farmers in 2019," *Markets Insider*, January 14, 2020, https://markets.businessinsider.com/news/stocks/trump-trade-war-impact-farmers-wisconsin-biggest-decline-on-record-2020-1-1028815780.

73 direct government payments to farmers in 2020: Dan Charles, "Farmers Got a Government Bailout in 2020, Even Those Who Didn't Need It," NPR, December 30, 2020, https://www.npr.org/2020/12/30/949329557/farmers-got-a-government-bailout-in-2020-even-those-who-didnt-need-it.

74 The most striking one was conducted by Jared McDonald: Jared McDonald, David Karol, and Lilliana Mason's survey on this subject was published in the journal *Political Behavior* in 2019 and entitled, winningly, "An Inherited Money Dude from Queens County: How Unseen Candidate Characteristics Affect Voter Perceptions." The researchers at some point in the proceedings offered accurate information to the respondents regarding how Trump's father influenced his son's career.

75 an heir accruing debt who liked nightclubs: David Barstow, Susanne Craig, and Russ Buettner, "Trump Engaged in Suspect Tax Schemes as He Reaped Riches from His Father," *New York Times*, October 2, 2018, https://www.nytimes.com/interactive/2018/10/02/us/politics/donald-trump-tax-schemes-fred-trump.html.

75 and Fred Trump had received financing: Thomas J. Campanella, "To the Manor Born: On the Rise of Fred C. Trump, Homebuilder," *Literary Hub*, October 3, 2019, https://lithub.com/to-the-manor-born-on-the-rise-of-fred-c-trump-homebuilder/.

77 her perceptive readings of the nonurban state of mind: The Democratic-leaning folks told Cramer that "pulling yourself up will only get you so far," while Republicans told her, "If only you work hard things will work out," and "the government is not the solution but the enemy."

78 a hearty number of union members also went for Trump in 2020: While Biden won 57 percent of union households nationwide, still 40 percent supported Trump. Ian Kullgren, "Union Workers Weren't a Lock for Biden. Here's Why That Matters," *Bloomberg Law*, November 10, 2020, https://news.bloomberglaw.com/daily-labor-report/union-workers-werent-a-lock-for-biden-heres-why-that-matters.

79 evaporating social status: "Loss aversion" was also present, I think, along with white nationalism and all sort of other terrible impulses on dark day of January 6, when rioters stormed the US Capitol: violent, mostly white crowds gathered to falsely claim Trump's presidential win. While some conspirators looked like stills from a hooligan version of a Wes Anderson movie, covered in Daniel Boone pelts, lost-cause-true-believers fighting "for our beloved President Donald J. Trump," they also included a West Virginia political delegate and a good number of the 82 seditious rioters who had been arrested by January 9 (the number grew to 440 by May) were middle-class, with fewer honest personal economic grievances than you might expect, including the CEO of a marketing consulting agency called Cogensia, who lived in a posh, suburban Midwest home and had enough money to donate $28,000 to Republican causes.

79 As the policy analyst Heather McGhee writes: Heather McGhee, "The Way Out of America's Zero-Sum Thinking on Race and Wealth," *New York Times*, February 13, 2021, https://www.nytimes.com/2021/02/13/opinion/race-economy-inequality-civil-rights.html?referringSource=artic leShare.

80 Today, two-thirds of American adults lack four-year degrees: Elise Gould, "Two-thirds of Adults Have Less Than a Four-Year Degree," Economic Policy Institute, June 22, 2018, https://www.epi.org/publication/two-thirds-of-adults-have-less-than-a-four-year-degree-policymakers-should-work-to-make-college-more-attainable-for-them-but-also-strengthen-labor-protections-that-help-all-workers/.

80 college degree: Philosopher Michael Sandel argues that Americans who are credentialed—with advanced or prestigious degrees—have bought into what Sandel calls "a rhetoric of rising." It's another false meritocracy, where high-end educations are twinned with worthiness rather than simply being "self-made" and without elite degrees. Perhaps those with credentials thus feel less than and ultimately seek recognition elsewhere, potentially becoming the "poorly educated" voters that Trump celebrated.

80 including earning less now than in 1979: "Real Wage Trends, 1979 to 2019," Congressional Research Service, updated December 28, 2020, https://fas.org/sgp/crs/misc/R45090.pdf.

81 many quarters of media: The people watching Fox and logging on to Breitbart were getting a very airbrushed rendition of Trump.

82 Trump's Latino voters in Florida and Nevada and elsewhere: Christian Paz,

"What Liberals Don't Understand About Pro-Trump Latinos," *Atlantic*, October 29, 2020, https://www.theatlantic.com/politics/archive/2020/10 /trump-latinos-biden-2020/616901/.

83   pieces in conservative publications during that time: Tanner Aliff, "Finding the Bootstraps My Father Always Talked About," *Daily Signal*, December 1, 2020, https://www.dailysignal.com/2020/12/01/finding-the -bootstraps-my-father-always-talked-about/.

83   the famed nineteenth-century Black writer Frederick Douglass: The now defunct far-right "anti-tyranny" publication the *Daily Fodder* also co-opted Douglass in 2020 for pro-gun purposes, writing that the abolitionist "understood the vital role firearms played in preserving individual freedom." I read these sites that reimagine Douglass as a John Wayne–style gunslinger and wince. Douglass knew firsthand that the institutions that were meant to be creating these independent Americans—from public schools to the ballot box to home ownership—were white dominated and racist by design.

84   that there were "no such men as self-made men": Douglass famously also wrote in this lecture, "We have all either begged, borrowed or stolen . . . ," highlighting our universal vulnerabilities and vices. I read him as ambivalent about the self-made man construction, as he also offers a somewhat positive view of the conceit in the same lecture, that the self-made men are "men of work."

Chapter 7: Zen Incorporated

87   mindfulness and meditation: Marianne Garvey, "Meditation Rooms Are the Hottest New Work Perk," *MarketWatch*, October 26, 2018, https:// www.marketwatch.com/story/meditation-rooms-are-the-hottest-new -work-perk-2018-10-26.

87   David Gelles, the *New York Times* business writer and a corporate-mindfulness guru: David Gelles, *Mindful Work: How Meditation Is Changing Business from the Inside Out* (New York: HarperCollins, 2015). This is according to research done by Aetna with Duke University in the early 2010s. As he writes in the book, "In 2012, as the mindfulness programs ramped up, health care costs fell 7 percent. That's $6.3 million going straight to the bottom line, partly thanks to mindfulness training, it appears."

87   placed meditation rooms throughout the company's San Francisco headquarters: Benioff described his decision, and the monks' input into the building's layout, at the 2016 *Forbes* CIO Summit. Alex Conrad, "How Monks Convinced Marc Benioff to Install 'Mindfulness Zones' Throughout Salesforce's New Offices," *Forbes*, March 7, 2016, https://www.forbes

.com/sites/alexkonrad/2016/03/07/how-monks-convinced-benioff-to-put
-mindfulness-zones-in-salesforce/?sh=479238ec1e24.

87 zealously suggested Transcendental Meditation to hundreds of his employees: Dalio introduced TM to his 735 employees around 2008. He has also donated tens of millions of dollars to the *Twin Peaks* director David Lynch's TM foundation. Richard Feloni, "The World's Largest Hedge Fund Reimburses Employees Half the Cost of $1,000 Meditation Lessons," *Business Insider*, November 10, 2016, https://www.businessinsider.com/transcendental-meditation-bridgewater-associates-2016-11.

88 more than half of Americans were profoundly unhappy in their jobs: I quote from the 2019 Great Jobs study conducted jointly by the Lumina Foundation, the Bill & Melinda Gates Foundation, Omidyar Network, and Gallup. These organizations surveyed 6,600 workers, and the 2020 Great Jobs study as well. "Not Just a Job: New Evidence on the Quality of Work in the United States," Gallup, October 20, 2019, https://www.gallup.com/education/267650/great-jobs-lumina-gates-omidyar-gallup-quality-download-report-2019.aspx.

88 a recent study of mindfulness programs: "For Mindfulness Programs, 'with Whom' May be More Important Than 'How,'" *News from Brown*, February 16, 2012, https://www.brown.edu/news/2021-02-16/mindfulness. Study author Brendan Cullen, working with Assistant Professor Willoughby Britton and her team, started their research looking at the effect of mindfulness practices on stress and depression, and learned that the group one practices with can also lead to emotional changes.

89 The meditation booths were called AmaZen: You can read more about these booths in articles like this piece in *Vice* titled "Amazon Introduces Tiny Zen Booths for Stressed Out Warehouse Workers," https://www.vice.com/en/article/wx5nmw/amazon-introduces-tiny-zenbooths-for-stressed-out-warehouse-workers; or *New York* magazine's "Amazon Amazen Booths Would Make Even a Dystopia Wince," by Sarah Jones, May 27, 2021, https://nymag.com/intelligencer/2021/05/amazon-amazen-booths-would-make-even-a-dystopia-wince.html.

90 strategies to keep us trapped by self-reliance myths: In a Fast Company round-up entitled "CEOs on Their Mindfulness Practices," Ryan Napierski, the president of the multilevel marking skin-care company Nu Skin, commented: "When I pray and meditate in the morning, I focus on that. When I mountain bike, I focus on the trail, so I don't crash. . . . These activities force me to have a single focus on what I'm doing right now. . . . So, when I am at the office, I can focus my efforts to be more productive and to be direct and decisive in my work." He was not interviewed,

of course, about what working for a multilevel marketing, or MLM, company might be doing to MLM workers having to sell Nu Skin (as anyone who has worked in selling products in this way can attest, a morning meditation isn't enough to make it any less stressful). All this is from Beck Bamberger, "8 CEOs on Their Mindfulness Practices," *Fast Company*, October 14, 2019, https://www.fastcompany.com/90416477/8-ceos-on-their-mindfulness-practices.

91 That was when Jon Kabat-Zinn promoted: Kabat-Zinn has written two bestselling books that have been translated into multiple languages, and offers DVD sets of three different meditation programs for sale through his website, https://www.mindfulnesscds.com/.

91 American Psychological Association's then-president Martin Seligman: Seligman said at the first Positive Psychology Summit in Lincoln, Nebraska (September 1999), "The most important thing, the most general thing I learned, was that psychology was half-baked, literally half-baked. We had baked the part about mental illness; we had baked the part about repair of damage. But as Corey was telling us today, that's only half of it. The other side's unbaked, the side of strength, the side of what we're good at."

93 "a trendy method for subduing employee unrest": Ron Purser and David Loy, "Beyond McMindfulness," *Huffpost*, July 1, 2013, https://www.huffpost.com/entry/beyond-mcmindfulness_b_3519289.

93 "Chewing on a raisin": Linnemann is a Dutch reporter

94 mindfulness was being marshaled: Sarah Perez, "Meditation and Mindfulness Apps Continue Their Surge Amid Pandemic," *TechCrunch*, May 28, 2020, https://techcrunch.com/2020/05/28/meditation-and-mindfulness-apps-continue-their-surge-amid-pandemic/.

94 five professors: Christopher Lyddy, Darren J. Good, Mark C. Bolino, Phillip S. Thompson, and John Paul Stephens, "Where Mindfulness Falls Short," *Harvard Business Review*, March 18, 2021, https://hbr.org/2021/03/where-mindfulness-falls-short

95 have questioned the concept: As the educational watchdog publication the *Hechinger Report* put it of "grit," with genteel dismissiveness, "criticisms range widely." This is what Jill Barshay wrote in her March 11, 2019, piece, "Research Scholars to Air problems with Using 'Grit' at School," https://hechingerreport.org/research-scholars-to-air-problems-with-using-grit-at-school/.

96 gratitude: Despite the tens of millions of gratitude hashtags and famous speakers like Tony Robbins proclaiming publicly about his own gratitude, posts by former "gratitude journal-ers" reveal trying to be grateful can

exact a high personal cost, with the state of being functioning as a kind of toxic positivity. As one such journal-er wrote, "My life was full of 'should bes' around gratitude in the absence of actual gratitude."

96 the ones who must adapt to the stressed workplace: David Forbes, *Mindfulness and Its Discontents: Education, Self, and Social Transformation* (Black Point, Nova Scotia: Fernwood, 2019).

97 teaching mindfulness to teenagers: Forbes discussed this framework in an October 2020 webinar presented by the Association for Contemplative Mind in Higher Education titled "Critical Social Mindfulness: Foundations and Emergent Practices for a New Mindful Deal."

Chapter 8: Go Fund Yourself

105 "dental fairs": These fairs were a phenomenon reported on by one of my organization's fellows, Bobbi Dempsey, who described her own front lower teeth as a jagged shard, as she, too, had grown up without the oral care she had needed.

106 He ultimately raised $4,015: Katie Kindelan, "8-Year-Old Presents School with $4K Check After Making Keychains to Pay off Classmates' Lunch Debts," *Good Morning America*, February 5, 2020, https://www.good morningamerica.com/living/story/year-presents-school-4k-check-making -keychains-pay-68725360.

108 survive on to begin with: Fauntleroy would have had a much easier time of it had he been born in France, Ireland, Germany, or Estonia, just to name a few countries that have superior social protections.

109 to continue to receive benefits: Aimee Picchi, "Social Security: Here's What Trump's Proposed Budget Could Mean for Your Benefits," *USA Today*, February 12, 2020, https://www.usatoday.com/story/money/20 20/02/12/social-security-trump-budget-aims-cuts-disabled-workers-pro gram/4738795002/.

109 in 2017, eleven million Americans said they had had "catastrophic medical expenses": Kristen Kendrick, "Despite ACA Coverage Gains, Millions Still Suffer 'Catastrophic' Health Care Costs," NPR, November 12, 2020, https://www.npr.org/sections/health-shots/2020/11/12/934146128 /despite-aca-coverage-gains-millions-still-suffer-catastrophic-health-care -costs.

110 as of September 2020, more than $625 million had been raised: Alison Van Houten, "GoFundMe: Enabling Giving," *TIME*, April 26, 2021, https://time.com/collection/time100-companies/5953754/gofundme -disruptors/.

110 640 colleges and universities operated food pantries on campus: As of May 2018, 640 food pantries were registered in the College & University Food

Bank Alliance (CUFBA), a national association, according to the CUFBA site, https://cufba.org/about-us/. The CUFBA site in 2021 says its numbers are now closer to seven hundred.

110 found that nearly half of them were food insecure: This was according to a study by the College & University Food Bank Alliance, the National Student Campaign Against Hunger and Homelessness, the Student Government Resource Center, and the Student Public Interest Research Groups partnered to conduct the survey between March and May 2016. The pandemic numbers came from Sara Goldrick-Rab et al., "#RealCollege During the Pandemic," the Hope Center, https://hope4college.com/wp-content /uploads/2020/06/Hopecenter_RealCollegeDuringthePandemic.pdf.

111 social media charity campaigns: As one character said in the film *Children of Men*, about a future of worldwide infertility where a few hardy souls try to save the human race on their own: "I can't really remember when I last had any hope, and I certainly can't remember when anyone else did either."

113 dystopian social safety net: In 2021, Barber created a campaign called a National Call for Moral Revival. He held events like one at the West Virginia Capitol to exert pressure on Senator Joe Manchin to finally support President Biden's bill that included crucial social services. Barber successfully enticed congresspeople to join him in his efforts and also used strategies like publishing digital ads in the state's newspapers to try to push Manchin.

Chapter 9: Mothers' Revolution

118 overlooked day care as a matter of course: Jordan Weissmann, "Child Care Is a Serious Economic Problem," Slate, February 11, 2019, https:// slate.com/business/2019/02/child-care-day-care-policies-paid-family -maternity-leave-gdp.html. Our scarce and pricey day care makes our country an anomaly globally.

120 the US Census Bureau revealed that in January 2021: This is according to research at the time by the Institute for Women's Policy Research. These out-of-work mothers can appear to resemble a social conservative's fantasy. As recently as 2013, a Pew poll showed that 51 percent of those surveyed believed that their kids were better off if their mother stays at home with them, while only 8 percent said that fathers shouldn't work. "Breadwinner Moms," Pew Research Center, May 29, 2013, https://www.pewresearch .org/social-trends/2013/05/29/breadwinner-moms/.

120 In 2020, nearly two in five caregivers: National Association for the Education of Young Children, "Holding On Until Help Comes: A Survey Reveals Child Care's Fight to Survive," July 13, 2020, https://www .naeyc.org/sites/default/files/globally-shared/downloads/PDFs/our-work

/public-policy-advocacy/holding_on_until_help_comes.survey_analysis
_july_2020.pdf. The survey lays out some more of these grim findings.

120 economist Katica Roy: Katica Roy, "It Won't Take 257 Years to Reach
Intersectional Gender Equity If We Do These 5 Things," *Fast Company*,
December 20, 2020, https://www.fastcompany.com/90588023/it-wont
-take-257-years-to-reach-intersectional-gender-equity-if-we-do-these-5-th
ings.

120 Rapid Assessment of Pandemic Impact on Development—Early Childhood Project: https://medium.com/rapid-ec-project.

122 Women were not only much more likely than men to lose their jobs or
be furloughed: Megan Cassella and Eleanor Mueller, "A Lack of Child
Care Is Keeping Women on Unemployment Rolls," *Politico*, June 25,
2020, https://www.politico.com/news/2020/06/25/child-care-women-un
employment-339012.

123 fathers have been found by researchers: Herminia Ibarra, Julia Gillard,
and Tomas Chamorro-Premuzic, "Why WFH Isn't Necessarily Good for
Women," *Harvard Business Review*, July 16, 2020, https://hbr.org/2020/07
/why-wfh-isnt-necessarily-good-for-women. The researchers found "that
women are more likely to carry out more domestic responsibilities while
working flexibly, whereas men are more likely to prioritize and expand
their work spheres."

123 "asymmetrical giving": As the philosopher Kate Manne wrote, women
"giving" to men is what our male-dominated society glides by on. Kate
Manne, *Down Girl: The Logic of Misogyny* (New York: Oxford University
Press, 2011).

125 "sourball of every revolution": Mierle Laderman Ukeles, "Manifesto for
Maintenance Art," 1969, https://queensmuseum.org/wp-content/uploads
/2016/04/Ukeles-Manifesto-for-Maintenance-Art-1969.pdf. I saw Ukeles's work with my curator friend Laura a few years ago at the Queens
Museum, including an installation of hundreds of gloves of sanitation
workers. I had never before seen art that was so reminiscent of my experience of mothering.

125 The title of Marvin Olasky's 1992 book *The Tragedy of American Compassion*: The book argued "for a biblical model for fighting poverty":
Daniel Bazikian, "Book Review: *The Tragedy of American Compassion*
by Marvin Olasky," Foundation for Economic Education, https://fee.org
/articles/book-review-the-tragedy-of-american-compassion-by-marvin
-olasky/.

126 it's less than 0.5 percent in the domestic product of the United States:
"Public Spending on Childcare and Early Education," OECD Family Da-
tabase, accessed August 20, 2021, https://www.oecd.org/els/soc/PF3_1

_Public_spending_on_childcare_and_early_education.pdf. It wasn't just the most indigent who were experiencing the hell of familial interdependence in a society that undervalues just that. "I'm one of many relatively rich people experiencing what poor people experience all the time—total abandonment by our government," as the *New York Times* columnist Michelle Goldberg put it. Band-Aids wouldn't resolve these structural inequalities.

127 "'global death cult'": Jedediah Britton-Purdy, "A Possible Majority," *Dissent*, October 27, 2020, https://www.dissentmagazine.org/online_articles/a-possible-majority.

129 I felt I always had to strive to survive: As Eula Biss writes in her limpid memoir, *Having and Being Had*, being middle class—part of a cadre of professionals and business owners—means that, like others of her ilk, she is defined as someone who makes investments, in real estate, in education, in the stock market, and also invests in the very idea of striving.

131 Swedish feminists lobbied zealously: Ingela K. Naumann, "Child Care and Feminism in West Germany and Sweden in the 1960s and 1970s," *Journal of European Social Policy*, February 1, 2005, https://journals.sagepub.com/doi/10.1177/0958928705049162.

Chapter 10: The Con of the Side Hustle

138 $39 billion by March 2021: Erin Griffith, "Instacart Raises $265 Million, More Than Doubling Its Valuation to $39 Billion," *New York Times*, March 2, 2021, https://www.nytimes.com/2021/03/02/technology/instacart-raises-265-million.html.

141 "These words have gained a strange kind of prestige from downwardly mobile, college-educated tech workers": John Patrick Leary, *Keywords: The New Language of Capitalism* (Chicago: Haymarket Books, 2019).

141 30 percent of Americans who do something else for pay in addition to their full-time jobs, according to an NPR/Marist survey: "A Close-Up Look at Contract Workers," NPR, January 22, 2018, https://www.npr.org/2018/01/22/579629353/a-close-up-look-at-contract-workers.

142 according to a survey conducted in 2020 by the Pew Research Center: Kim Parker, Juliana Menasce Horowitz, and Anna Brown, "About Half of Lower-Income Americans Report Household Job or Wage Loss Due to COVID-19," Pew Research Center, April 21, 2020, https://www.pewresearch.org/social-trends/2020/04/21/about-half-of-lower-income-americans-report-household-job-or-wage-loss-due-to-covid-19/#many-adults-have-rainy-day-funds-but-shares-differ-widely-by-race-education-and-income.

142 "a knot in your stomach, a rash on your skin, are losing sleep, and losing touch with your wife and kids": Nitasha Tiku, "The Gospel of Hard

Work, According to Silicon Valley," *Wired*, June 2017, https://www.wired.com/2017/06/silicon-valley-still-doesnt-care-work-life-balance/.

143 scholar Tressie McMillan Cottom observes: Tressie McMillan Cottom, "The Hustle Economy," *Dissent*, Fall 2020, https://www.dissentmagazine.org/article/the-hustle-economy.

144 additional job: As Malcolm X wrote, "everyone in Harlem needed some kind of hustle to survive," which could mean anything from running the numbers to a side business.

146 Meanwhile, the Uber "Greenlight Hubs": Kia Kokalitcheva, "Uber Temporarily Closes Local Hubs for Drivers amid Virus Outbreak," Axios, March 13, 2020, https://www.axios.com/uber-temporarily-closes-local-hubs-for-drivers-amid-virus-outbreak-a6778d15-5f56-465a-bff8-cb3de4b164d4.html.

147 "change as we find it necessary to change it, as we go on making our own language and history," according to Raymond Williams: *Keywords: A Vocabulary of Culture and Society* (Kent, UK: Croom Helm, 1976).

147 the word *security* when it comes to work: Chuck Marr et al., "House COVID Relief Bill Includes Critical Expansions of Child Tax Credit and EITC," Center on Budget and Policy Priorities, March 2, 2021, https://www.cbpp.org/research/federal-tax/house-covid-relief-bill-includes-critical-expansions-of-child-tax-credit-and.

148 no other source of support: The United Food and Commercial Workers Union Local 1546 in Chicago, which represents grocery workers, criticized Instacart for these layoffs, which included ten of its own members.

149 "small parcels of nervous energy picked up by the recombining machine": Franco Bifo Berardi, *After the Future* (Chico, CA: AK Press, 2011).

Chapter 11: Class Traitors

155 "willing to do the only thing that is actually going to get us there": Juliana Kaplan, "Inequality Flamethrower Anand Giridharadas on Why Billionaires Shouldn't Exist and His Hopes for the Biden Administration," *Business Insider*, December 8, 2020, https://www.businessinsider.com/anand-giridharadas-why-billionaires-shouldnt-exist-biden-philanthropy-inequality-2020-12.

156 their group overall is a tiny minority: The wealthy are not just a minority in America, they are a tiny segment of the population. They're also getting richer at a staggering rate. During the 2020 COVID-19 pandemic, 614 billionaires increased their net worth by nearly $1 trillion. Aimee Picchi, "U.S. Billionaires Gained Almost $1 Trillion in Wealth During the Pandemic," *CBS News*, October 20, 2020, https://www.cbsnews.com/news/billionaires-pandemic-1-trillion-wealth-gain/.

156 but they do, demographically, put Ho in the top 10 percent: "Are You Rich? Where Does Your Net Worth Rank in America?," *New York Times*, August 12, 2019, https://www.nytimes.com/interactive/2019/08/12/upshot/are-you-rich-where-does-your-net-worth-rank-wealth.html.

158 dubious loopholes like carried interest: The Tax Policy Center defines the carried interest loophole as "Carried interest, income flowing to the general partner of a private investment fund, often is treated as capital gains for the purposes of taxation." "Briefing Book," Tax Policy Center, accessed January 2021, https://www.taxpolicycenter.org/briefing-book/what-carried-interest-and-how-it-taxed.

158 Warren's own proposal: Emmanuel Saez and Gabriel Zucman, "Progressive Wealth Taxation," BPEA Conference Draft, Brookings, Fall 2019, https://www.brookings.edu/bpea-articles/progressive-wealth-taxation/. Also by Saez and Zucman: "Jobs Aren't Being Destroyed This Fast Elsewhere. Why Is That?," *New York Times*, March 30, 2020, https://www.nytimes.com/2020/03/30/opinion/coronavirus-economy-saez-zucman.html.

159 "designed to counter unsustainable behaviors": Allison Christians and Tarcisio Diniz Magalhaes, "The Case for a Sustainable Excess Profits Tax," SSRN, March 24, 2021, https://ssrn.com/abstract=3811709.

160 Schoenberg had posted portions of his returns online: Eric Schoenberg, "How I Paid Only 1% of My Income in Federal Income Tax," *Huffpost*, April 25, 2011, https://www.huffpost.com/entry/how-i-paid-1-of-my-income_b_852948.

160 wealth generated in the prior year went to the richest 1 percent: "'World's Richest 1% Get 82% of the Wealth,' says Oxfam," Katie Hope, *BBC News*, January 22, 2018, https://www.bbc.com/news/business-42745853. This is global as well. As UNICEF had found, there is a sickeningly unequal distribution of wealth: UNICEF, "1 in 5 Children in Rich Countries Lives in Relative Income Poverty, 1 in 8 Faces Food Insecurity," June 15, 2017, https://www.unicef.org/press-releases/1-5-children-rich-countries-lives-relative-income-poverty-1-8-faces-food-insecurity.

160 As the near-billionaire philanthropist Nick Hanauer has written: Yoni Appelbaum, "Is Big Philanthropy Compatible with Democracy?," *Atlantic*, June 28, 2017, https://www.theatlantic.com/business/archive/2017/06/is-philanthrophy-compatible-democracy/531930/.

161 palliate a teensy-weensy piece of student debt overall: "A Look at the Shocking Student Loan Debt Statistics for 2022," Student Loan Hero, February 7, 2022, https://studentloanhero.com/student-loan-debt-statistics/#:~:text=General%20student%20loan%20debt%20facts&text=The%20most%20recent%20data%20indicate,Americans%20with%20student%20loan%20debt.

163 according to the Council for Aid to Education: Emily Arnim, "How Community Colleges Attract Major Gifts," August 26, 2021, EAB, https://eab .com/insights/daily-briefing/advancement/how-community-colleges-are -attracting-major-gifts/.

Chapter 12: The Feeling Is Mutual Aid

165 For months, New York City's residents: The network of community refrigerators has continued to grow in New York and beyond, and the impact of one on the community of Ridgewood, Queens, led local car service owner Antonio Vaca to start a free food market. Jessy Edwards, "Inside the Brooklyn Fridge That Started a Mutual Aid Movement," *BKReader*, April 27, 2021, https://www.bkreader.com/2021/04/27/inside-the-brook lyn-fridge-that-started-a-mutual-aid-movement/?fbclid=IwAR1XNF h6VqIrRmVuGJHFqOZq_fPDqosgOL4_NIhkTv-wpqOpUywf-C e1Kn0.

166 In Columbia, South Carolina, Dylan Gunnels, who founded the mutual aid group the Agape Table: "Dylan Gunnels Interview," COVID-19 Oral History Class Collection, May 19, 2020, https://digital.library.sc.edu /exhibits/covid-oral-history/interviews/dylan-gunnels-interview/.

166 "Why Today's Social Revolutions": Rinku Sen, "Why Today's Social Revolutions Include Kale, Medical Care and Help with Rent," Zócalo Public Square, July 1, 2020, https://www.zocalopublicsquare.org/2020/07/01 /mutual-aid-societies-self-determination-pandemic-community-organ izing/ideas/essay/.

166 Anke al-Bataineh created a mutual aid network of neighbors: Jean Hopfensperger, "In Minnesota, Mutual Aid Groups Surge in Wake of Floyd Death," *Minneapolis Star Tribune*, August 10, 2020, https://www.star tribune.com/in-minnesota-mutual-aid-groups-surge-in-wake-of-floyd -death/572060882/.

167 organized under the ancient Greek name Lysistrata: "We Are an Online, US Based, Community Run Sex Worker Activist Cooperative," Lysistrata, accessed February 14, 2022, https://www.lysistratamccf.org.

167 "We can buy into the old frameworks": Jia Tolentino, "What Mutual Aid Can Do During a Pandemic," *New Yorker*, May 11, 2020, https://www .newyorker.com/magazine/2020/05/18/what-mutual-aid-can-do-during -a-pandemic.

171 Czar Alexander II's most beloved teen assistant: Peter Kropotkin, *Mutual Aid: A Factor of Evolution* (London: Freedom Press, 2009, reprint).

172 It also *came* from Darwin: Charles Darwin, *The Descent of Man, and Selection in Relation to Sex* (London: J. Murray, 1871).

173   "If Kropotkin overemphasized mutual aid": Stephen Jay Gould, *Natural History* 97, no. 7 (1988): 12–21.

173   spent a year studying bonobos: Johnson saw the bonobos fighting, but over the course of an hour they would inevitably make up and become one with the group, including Johnson himself; he believed they thought of him as a "the strange hairless ape."

174   Darwin's "pro-social" view of nature: Jason G. Goldman, "7 Questions with . . . Eric M. Johnson," *Scientific American*, September 3, 2010, https://blogs.scientificamerican.com/thoughtful-animal/7-questions-with-8230 -eric-m-johnson.

174   community-minded version of Darwin: Of course, we mustn't forget that critics find evidence of "scientific racism" in his work, citing examples like this sentence from *The Descent of Man*: that at "some time the civilized races of man will exterminate and replace throughout the world the savage races."

174   the late scholar and activist David Graeber wondered: David Graeber, "What's the Point If We Can't Have Fun?," *The Baffler* 24 (January 2014), https://thebaffler.com/salvos/whats-the-point-if-we-cant-have-fun.

174   "A Symbiotic View of Life": In an essay by Scott Gilbert, Jan Sapp, and Alfred I. Tauber from 2012 titled "A Symbiotic View of Life: We Have Never Been Individuals," the authors write that while "the notion of the 'biological individual' is crucial to studies of genetics, immunology, evolution . . ." in the twenty-first century, the level at which animals and plants are interconnected with "symbiotic microorganisms" has come to disrupt the boundaries of the individual, https://www.journals.uchicago.edu/doi/full/10.1086/668166.

175   Some of the responses to the phrase "mutual aid": A man with the Twitter handle "staying in (on the lam)" tweeted, "Still blows my mind that people will so willingly give to a national aid or some random non-profit because it has a dot org but they won't give even $5 to a Venmo/PayPal Mutual Aid . . ." Or "Puff the Magic Hater," who wrote, "pls Don't turn your mutual aid projects into nonprofits."

176   Kristin Ross calls "communal luxury": Kristin Ross, *Communal Luxury: The Political Imaginary of the Paris Commune* (Brooklyn, NY: Verso, 2016).

Chapter 13: Boss Workers

180   higher than at other businesses—$19.67 per hour: Tim Palmer, "2019 Worker Cooperative State of the Sector Report," Democracy at Work Institute, January 29, 2020, https://institute.coop/resources/2019-worker -cooperative-state-sector-report.

181 rolled out the Drivers Cooperative: The precursor to the Drivers Cooperative could be the Union Cab Collective, founded in Madison, Wisconsin, in 1979, which proved that worker/driver cooperatives could function and thrive. Today, Union Cab has 157 members, 109 of whom are primarily drivers. It started taking online orders in the 1990s and launched its own app in 2017.

186 a national network of cooperative groups: Du Bois published a monograph in 1907 as part of a series titled "Economic Cooperation Among Negro Americans." At the time, he listed 154 African American–owned cooperative businesses: 14 "producer cooperatives," 3 "transportation cooperatives," 103 "distribution or consumer cooperatives," and 34 "real estate and credit cooperatives."

187 as much as their white counterparts in that time: Bruce J. Reynolds, "Black Farmers in America, 1865–2000: The Pursuit of Independent Farming and the Role of Cooperatives," USDA, RBS Research Report, October 2003, https://www.rd.usda.gov/files/RR194.pdf.

187 The third sprouting of the Black cooperatives: If I were a filmmaker, I'd try to make a film like *Hidden Figures*, but about Black cooperatives in the 1960s. *Hidden Figures*, if you recall, was an inspiring albeit schmaltzy Hollywood pic that documented the contributions of a group of Black NASA female mathematicians who were central to the space race. Instead of making such a film, for now I'm writing this chapter.

189 were responsible for the unhealthy conditions: Albert Samaha and Katie J. M. Baker, "Smithfield Foods Is Blaming 'Living Circumstances in Certain Cultures' for One of America's Largest COVID-19 Clusters," *Buzzfeed*, April 20, 2020, https://www.buzzfeednews.com/article/albert samaha/smithfield-foods-coronavirus-outbreak.

Chapter 14: Inequality Therapy

192 physical and sexual abuse: Christine Cissy White, #BeReal—Christine Cissy White," *Hasty Words* (blog), August 7, 2015, https://hastywords. com/2015/08/07/bereal-christine-cissy-white/ Accessed on March 4, 2022.

193 the original ACEs Study, the CDC-Kaiser Permanente Adverse Childhood Experiences Study: this was a huge investigation into childhood abuse and neglect that ran from 1995 to 1997, collecting data from seventeen thousand Health Maintenance Organization members, https://www.cdc.gov /violenceprevention/aces/about.html.

195 intersection of emotional suffering and inequity: These are forms of therapy and counseling that combine social awareness and collective identity with more traditional therapeutic goals like individual healing.

196 Harriet Fraad: "Dr. Harriet Fraad," accessed January 25, 2021.

198 the competitive nature of today's market system may be at the root of many clients' problems: Kalman Glantz and J. Gary Bernhard, *Self-Evaluation and Psychotherapy in the Market System* (Oxfordshire, UK: Routledge, 2018).

198 a 2020 study of US adults ages thirty and over: J. M. Twenge and A. B. Cooper, "The Expanding Class Divide in Happiness in the United States, 1972–2016," *Emotion* (2020, advance online publication), https://doi.org /10.1037/emo0000774.

199 "statistically significant positive relationship between income inequality and risk of depression": Vikram Patel, Jonathan K. Burns, Monisha Dhingra, Leslie Tarver, Brandon A. Kohrt, and Crick Lund, "Income Inequality and Depression: a Systematic Review and Meta-Analysis of the Association and a Scoping Review of Mechanisms," *World Psychiatry*, February 17, 2018: 76–89, https://www.ncbi.nlm.nih.gov/pmc/articles/PMC 5775138/.

199 as if they were already in some way suspect: Thomas E. Trail and Benjamin R. Karney, "What's (Not) Wrong with Low-Income Marriages," *Journal of Marriage and Family* 73, no. 3 (May 24, 2012): 413–27, https://online library.wiley.com/doi/abs/10.1111/j.1741-3737.2012.00977.x.

199 a 2020 analysis by Princeton University: Nathan N. Cheek and Eldar Shafir, "The Thick Skin Bias in Judgments About People in Poverty," *Behavioural Public Policy*, Cambridge University Press, August 14, 2020: 1–26, https://www.cambridge.org/core/journals/behavioural-public-policy /article/thick-skin-bias-in-judgments-about-people-in-poverty/2A8CCE1 3402F69C2B0D1145BE5270E1D.

200 the American Sociological Association's *Journal of Health and Social Behavior*: Vikram Patel et al., "Income Inequality and Depression: A Systematic Review and Meta-Analysis of the Association and a Scoping Review of Mechanisms," *World Psychiatry* 17, no. 1 (February 2018): 76–89, https:// www.ncbi.nlm.nih.gov/pmc/articles/PMC5775138/.

Chapter 15: Volunteering Ourselves

209 our volunteerism also has a dark side: Nina Eliasoph, *Making Volunteers: Civic Life at Welfare's End* (Princeton, NJ: Princeton University Press, 2011).

213 Celina Su surveyed twenty-five thousand PB participants: Celina Su: "Beyond Inclusion: Critical Race Theory and Participatory Budgeting," *New Political Science* 39, no. 1 (2017) 126–42, https://www.tandfonline .com/doi/full/10.1080/07393148.2017.1278858.

214 according to Alexander Kolokotronis, a young scholar: Alexander Kolokotronis, "What to Do Once We've Defunded the Police," *Current Affairs*,

July 9, 2020, https://www.currentaffairs.org/2020/07/what-to-do-once
-weve-defunded-the-police.

Chapter 16: Unmaking the Self-Made Myth

218   hybrid narratives: This sort of narrative was proposed as the most effective
format in the Norman Lear Center's March 2021 report "Stories Matter"
by Erica L. Rosenthal. While "personal responsibility narratives" in films
or television shows "focus on individual choices and responsibility" with
an "emphasis on willpower or lifestyle choices," "hybrid narratives" combine "personal responsibility with external factors, situating individual
stories within a larger structural context. . . . Research has shown that
audiences experience greater empathy in response to this type of narrative."

220   by August 2021, America's billionaires could have covered a $3,400
check: They could have but they didn't, like bullies who took everyone
else's sweets—as a kind of "moral injury." What do I mean by that? Psychiatrists tend to use "moral injury" now to describe the trauma that, say,
soldiers experience after battles or killing—when they enact or fail to prevent or simply see events that contradict their fundamental beliefs. But I'd
argue that people who have survived the pandemic economically intact experience a kind of "moral injury" when so many Americans were threatened to be ejected from their homes.

221   of other organizations: These included Academy of American Poets and
PEN America.

221   "media extinction event": In 2020, thirty-seven thousand media workers
had lost their jobs—poof! One of the many who came to me and my organization for financial help was a fifty-one-year-old photojournalist who
four different doctors agreed had coronavirus and was suddenly unable to
pay their rent. They needed money—and we gave them what we could,
$1,500, to continue to stay in the profession and pay their rent.

222   1933 painting: I visited this Alice Neel painting during the pandemic, the
1933 work *Investigation of Poverty*, at the Russell Sage Foundation, which
showed two impoverished old men on one side and foundation workers on
the other, the latter staring in an apathetic-seeming way at the poor guys
they were supposed to be helping. (The painting has little to do with today's venerable Russell Sage Foundation but rather that entity nearly a
century ago.) I didn't think it was happenstance that Neel's own painful familial experiences gave her insight into the perils of detached philanthropy.

223   You will likely have your own list: One inspiration for this practice comes
from the podcaster and author Bob McKinnon, the creator of Your Amer-

ican Dream Score, who encourages guests on his podcast *Attribution* to credit those who have helped them for this very reason.

223 Michelle Hughes at the National Young Farmers Coalition: Michelle Hughes, "Dismantling the Bootstrap Myth," ExtraNewsfeed, January 16, 2017, https://extranewsfeed.com/dismantling-the-bootstrap-myth-8ba 295f92431. As Hughes writes, "Simply put, we live in a nation in which some are issued boots and some are not."

225 We might embrace the notion of "secular faith": Martin Hägglund writes in his book *This Life: Secular Faith and Spiritual Freedom*, "Both capitalism and religion prevent us from recognizing in practice that our own lives—our only lives—are taken away from us when our time is taken away from us."

225 "new main characters, better plots, and at least the possibility of some happy endings," as one activist writer puts it: That writer is Lynne Segal in her book *Radical Happiness: Moments of Collective Joy* (Brooklyn, NY: Verso, 2017).

225 radiant with expressiveness: Vivian Gornick, *The Romance of American Communism* (Brooklyn, NY: Verso, 2020).

Epilogue: My Interdependence Day

228 bestselling illustrated memoir: I am thinking of a drawing from the artist Alison Bechdel's 2021 book *The Secret to Superhuman Strength*, in which the artist conceives of her own participation in workout crazes and her compulsive exercise as a kind of twisted Emersonian striving.

230 Marilyn Friedman, "relational autonomy": And as law and political science professor Jennifer Nedelsky writes in *Law's Relations: A Relational Theory of Self, Autonomy and Law*, "People are not simply self-made," rather they are constituted socially.

Index